Best wishes
from the author

Мдшилевич

Vibroacoustical Diagnostics for Machines and Structures

MECHANICAL ENGINEERING RESEARCH STUDIES

ENGINEERING DYNAMICS SERIES

Series Editor: **Professor J. B. Roberts,** *University of Sussex, England*

1. Feedback Design of Systems with Significant Uncertainty*
   M. J. Ashworth

2. Modal Testing: Theory and Practice
   D. J. Ewins

3. Transfer Function Techniques and Fault Location*
   J. Hywel Williams

4. Parametric Random Vibration
   R. A. Ibrahim

5. Statistical Dynamics of Nonlinear and Time-Varying Systems
   M. F. Dimentberg

6. Strategies for Structural Dynamic Modification
   John A. Brandon

7. Fragility Analysis of Complex Structural Systems
   F. Casciati *and* **L. Faravelli**

8. Vibroacoustical Diagnostics for Machines and Structures
   M. F. Dimentberg, K. V. Frolov *and* **A. I. Menyailov**

* out of print

Vibroacoustical Diagnostics for Machines and Structures

M. F. Dimentberg
K. V. Frolov
and
A. I. Menyailov

USSR Academy of Sciences, Moscow

RESEARCH STUDIES PRESS LTD.
Taunton, Somerset, England

JOHN WILEY & SONS INC.
New York · Chichester · Toronto · Brisbane · Singapore

RESEARCH STUDIES PRESS LTD.
24 Belvedere Road, Taunton, Somerset, England TA1 1HD

Marketing and Distribution:

Australia and New Zealand:
Jacaranda Wiley Ltd.
GPO Box 859, Brisbane, Queensland 4001, Australia

Canada:
JOHN WILEY & SONS CANADA LIMITED
22 Worcester Road, Rexdale, Ontario, Canada

Europe, Africa, Middle East and Japan:
JOHN WILEY & SONS LIMITED
Baffins Lane, Chichester, West Sussex, England

North and South America:
JOHN WILEY & SONS INC.
605 Third Avenue, New York, NY 10158, USA

South East Asia:
JOHN WILEY & SONS (SEA) PTE LTD.
37 Jalan Pemimpin 05-04
Block B Union Industrial Building, Singapore 2057

Library of Congress Cataloging-in-Publication Data

Dimentberg, M. F. (Mikhail Fedorovich)
Vibroacoustical diagnostics for machines and structures / M. F. Dimentberg, K. V. Frolov, and A. I. Menyailov.
p. cm.—(Mechanical engineering research studies) (Engineering dynamics series; 8)
Includes bibliographical references and index.
ISBN 0-86380-116-1 (Research Studies Press).—ISBN 0-471-93004-0 (Wiley)
1. Vibration tests. 2. Structural dynamics. 3. Machinery—Vibration—Testing. 4. Acoustic emission testing. I. Frolov, K. V. II. Menyailov, A. I. (Anatolyi I.) III. Title. IV. Series.
V. Series: Engineering dynamics series; 8.
TA355.D487 1991
620.3'7—dc20 90-29896
CIP

British Library Cataloguing in Publication Data

Dimentberg, M. F.
Vibroacoustical diagnostics for machines and structures.—(Mechanical engineering research studies—engineering dynamics series)
I. Title II. Frolov, K. V. III. Menyailov, A. I.
IV. Series
531

ISBN 0 86380 116 1

ISBN 0 86380 116 1 (Research Studies Press Ltd.)
ISBN 0 471 93004 0 (John Wiley & Sons Inc.)

Printed in Great Britain by Galliard (Printers) Ltd., Great Yarmouth

Editorial Foreword

High amplitude vibration in structural and mechanical systems is usually very undesirable. At worst it can lead to mechanical failure, usually through fatigue. Even if the system does not actually fail, severe degradation of performance can occur through, for example, the wearing or fretting of components. In addition, a vibrating system can transmit unacceptably high levels of noise to its environment, or cause excessive vibration in neighbouring structures and machines. It is important, therefore, to ensure, at the design stage, that vibration amplitudes will be at acceptably low levels.

In correctly designed systems the vibration behaviour, under normal operating conditions, can be measured fairly easily, even when the system is within a safe regime. Measured 'vibration signatures' contain a wealth of information which can be used to monitor 'health' — that is, the extent to which the system is performing satisfactorily. Thus, changes in the characteristics of the vibration signature can be used for a variety of diagnostic purposes. For example, they can be used to give an early indication of a malfunction, providing the opportunity for action to be taken before serious damage occurs. However, for the detection of significant changes, some of which may be quite subtle, one needs high speed data capture facilities, together with appropriate sophisticated data processing algorithms.

Fortunately, the extraordinary progress in computer technology in recent years has enabled the development of cheap, yet powerful, data capture systems. This has provided the impetus for much research into software-based diagnostic techniques, over the last two or three decades. The authors have brought together much of their own work in this area, and that of others, in a book which is thorough and yet highly readable. It is based on a wealth of experience which they have gained through consultative work with industry. By concentrating on physical aspects, and keeping the mathematical details to an absolute minimum, they have succeeded in presenting the basic concepts in an exceptionally clear and simple manner.

This is a very timely and important contribution to a subject which will continue to grow in importance, in a wide range of engineering, in the foreseeable future.

J.B. Roberts
July 1991

Contents

Preface

This book is dedicated to vibroacoustical (VA) diagnostics of machines and structures; that is, to the estimation or monitoring of their internal state through an appropriate processing of measured vibrational and/or noise signals. In general, this topic has quite a long history. As far back as in the twenties and thirties many experienced people could be found, in various industries, who were able to make assessments of a machine or structure just by careful listening to its sound emissions during operation. (In fact, the first specialist in VA diagnostics was probably a person from the Stone Age who discovered that, if it became impossible to hear the heart-beat of his wounded fellow-tribesman, then he was dead, whereas an audible heart-beat implied that he was still alive.)

However, real successes have come to VA diagnostics only in the last two or three decades, with the advent of sophisticated hardware and software for signal processing, together with more sensitive measurement equipment. In particular, certain newly developed fields of applied mathematics, such as statistical dynamics and pattern-recognition theory, which have been successfully applied in radar and sonar technologies, have also become increasingly useful for VA diagnostics. Specifically, they form the basis for certain algorithms for VA diagnostics; that is, for those procedures of VA signal processing that may provide useful information on the internal state of a machine, mechanism, structure etc. Nowadays, a VA diagnostic system may comprise an essential part of a machine or structural system. The availability of such a system for a nuclear power plant, for example, strongly improves its potential for safety and reliability.

Internal-state monitoring of a machine or structure may be desirable both in the course of its operation and at the design stage, when the first prototype is tested. The aims of VA diagnostics in these two cases are somewhat different. Whereas in the former the principal aim may be to detect the occurrence of possible faults in a properly designed machine (structure), in the latter the principal objective may be to decide if any design modifications and/or corrections are necessary, or advantageous. Moreover, the corresponding algorithms of VA diagnostics in these two cases may be quite different. Firstly, this is because of the different requirements of the diagnostic systems, reflecting their different aims. The other reason for these differences is the fact that, for a normally operating machine (structure), VA signal measurements may start at the very beginning of its service life. Therefore, with this 'normal' VA signal available, its changes in the course of the service life may be used with advantage in algorithms for the detection of changes within the initially normal machine (structure). On the other hand, whenever the prototype is tested for the first time, such a baseline, or reference VA signal, may not be available (more exactly, the measured signal may not yet be classified at this stage as being the 'normal' one). In this case, the diagnostic system must rely only on an analysis of the VA signal for the current state of the machine or structure; therefore, a so-called 'identificational' approach to diagnostics, which will be explained later, becomes crucial.

In this book, certain basic ideas concerning VA diagnostics and their implementation are presented, together with specific examples from various fields of engineering. The book has one very specific feature. Specifically, although these basic ideas and algorithms are of mathematical origin, and of a mathematical nature, an attempt is made here to provide a presentation in a form suitable for a fairly broad category of readers — both for those who are, and for those who are not, especially familiar with mathematics (the reader is presumed to have only a basic knowledge, such as that normally gained in secondary-school programmes). Therefore, potential readers may include managers, maintenance staff of plants etc. who wish to know in general terms how diagnostic systems operate, without going into mathematical details and technicalities. Moreover, the book is simple enough, in general, to be suitable for nonspecialists who are interested in this field of science and technology. On the other hand, engineers and research scientists with a sufficient mathematical background will also find in this book something of interest, together with references to relevant specialist literature, where more detailed mathematical derivations and/or descriptions of the results may be found.

The challenging task of presenting the material with as little mathematics as possible is attempted here by introducing the basic concepts of statistical dynamics, signal processing etc. in physical terms. The same style is preserved in the description of the various algorithms of VA diagnostics. These algorithms are presented as a rule without full derivations but with detailed descriptions of their physical meaning and implications. Certain purely mathematical aspects of some procedures for signal processing are also omitted, with due reference to relevant literature. In short, by representing a VA diagnostic system by a flow chart, together with the mathematical model of the VA signal generation within a machine or structure, it is hoped that the reader will get a clear understanding of the properties of the input and output of each block. The operation of these blocks is outlined sometimes only in qualitative terms; it seems, however, that this is just the type of description that is needed by many people involved in VA diagnostics, or mechanical signature analysis.

The distribution of the material between the various chapters is mostly made according to the basic general problems and/or methods of diagnostics or identification, rather than according to specific fields of application. On the other hand, examples of the application of these general approaches are related to various industries, possibly with a slight preference for nuclear power engineering*. Such an interdisciplinary approach to the presentation is useful in providing a good general insight into the behaviour of various classes of machines and structures (from the point of view of formulating diagnostic criteria and rules) as well as for an exchange of VA diagnostic techniques between various fields of application. While some readers may be slightly disappointed by the lack of direct answers to their specific problems and questions, the authors hope that this insight, together with an acquaintance of experiences from other fields, will compensate for this lack and provide some additional inspiration.

---

* By the way, the authors are fully aware of the possible ecological impact of the nuclear power industry on the environment. However, being equally aware of the fact that further progress in this field is inevitable at present, in view of the limited availability of natural energy resources (at the very least, it seems impossible to shut down all existing plants immediately), they decided that it would be proper to make a contribution, within their powers, concerning further means of improving the reliability and safety of nuclear power plants

Introduction

One of the authors of this book is a great fan of coffee. Every morning he starts preparing his coffee by grinding coffee grains in an electric mill; so he knows very well how the noise of the coffee mill changes during the course of the grinding process. At first, various fairly low-frequency components are dominant in this noise. Then, presumably after the coffee particles have become rather small, the mill noise approaches a so-called pure tone, or a sinusoidal sound. The frequency of the latter gradually increases, and finally seems to reach the rotational frequency of the mill shaft, i.e. 50 Hz or cycles per second (the frequency of the commercial alternating current in Europe).

A specialist in physics or mechanics may readily provide an explanation for these phenomena based on a qualitative, and maybe even partly quantitative, analysis of the noise due to the crushing of large-sized grains and through a consideration of the variation of the particles' resistance to the shaft rotation as they decrease in size. Such a model could, in principle, prove to be of some use to the designers of coffee mills.

However, for the user of the mill, who is, in effect, 'a member of the maintenance staff', this physical model is not especially important as long as the mill operates normally. He wishes to know only one thing — whether the coffee particles have become sufficiently fine for the mill to be switched off. In fact, the present author may claim, with his long-term experience as such a user, that he can discriminate confidently between the noises of the mill with fine, and not fine enough, coffee particles.

This example from everyday experience illustrates clearly one of the possible approaches to VA diagnostics; it is based on so-called empirical pattern recognition. The discrimination between different types of machine noise according to this approach (in this example, between noise signals for 'ready' and 'not yet ready' states) is based solely on previous experience, rather than on an analysis of the process of the VA signal generation. The fact that, in this example, the discrimination is made by a human being, rather than by a computer, need not be emphasised here. We consider rather the following topic: how can such a discrimination be made automatically by a computer or, more precisely, what is the possible algorithm for such a discrimination; that is, the procedure for noise-signal processing, with subsequent proper decision-making based on properly processed signals?

It should be added only that, so far as such empirical methods of diagnostics on the basis of previous experience are concerned, it may sometimes prove to be rather difficult for such an automatic computer-based diagnostic system to compete successfully with a human operator. A good and clear example is that of a car mechanic, or an experienced car user, who can reliably detect extraneous or 'abnormal' sounds, say, in the noise signals from a car engine and/or suspension during a ride, and can identify the origin of such sounds. The basis for the success of such VA diagnostics by a human being lies not only in the extremely high sensitivity of the human ear, but also in the specific recognition algorithms which are involved. Although these 'human' algorithms for learning and decision-making are not yet completely understood, insight into such algorithms may be used with advantage in developing software for computer-based diagnostic systems. In any case, however, the latter may, in certain respects, be inferior, compared with an experienced human operator. For example, computer programs for playing chess cannot (yet!) beat the best human chess players.

The answer to the problem of formulating suitable automated signal processing software for a computer may be provided by the so-called theory of pattern recognition [24]. This theory deals with a finite number of variables. The latter may be regarded as comprising a multidimensional space, or state space, where each variable corresponds to a certain axis in this space and each combination of fixed values of the variables corresponds to a certain point in this space. Let us assume now that the state space is divided into several domains and that information on some points is fed into a computer; that is, their coordinates are stored together with a statement showing to which domain each of these points belongs. Then, on increasing the number of these given points from known domains, the computer may actually gradually 'learn' to classify the subsequent points; that is, to evaluate their domains simply from the given coordinates. This can be done by well established procedures for the approximate reconstruction of borderlines between different domains from known data on a finite number of points in the state space. These procedures, which are based, in fact, on some interpolation of the borderlines, are used widely, for example, in the classification of images (which should be properly quantified, of course). Thus, if, say, 1000 images of cats and 1000 images of dogs are 'shown' to the computer, then the latter would be able to recognise the 2001th image as being that of either a cat or a dog. (It should be noted, though, that such a classification would be correct only within a given set of domains; thus, if the computer 'sees' a rabbit or a crocodile, it will 'correctly' classify it also as being either a cat or a dog.)

Thus, it can be seen that the success of such an approach to VA diagnostics depends crucially on the availability of sufficient VA signal data for all possible states of the machine or structure; for example, both for the normal state as well as for the states with all possible or probable faults, if fault detection is the principal aim of the diagnostic system. In the above example of a coffee mill, this is definitely not a problem. In a fault-detection problem, difficulties may arise both with foreseeing all the possible types of fault and with getting the necessary information on the VA signals in the faulty states; in fact, operation of certain machines and structures in such states may be completely prohibitive. A clear example is that of an aircraft in a state of wing flutter (see Chapter 7). Nevertheless, this direct empirical pattern-recognition approach to VA diagnostics should be regarded in general as a progressive one, its areas of applicability gradually expanding with the accumulation of measured data for operating machines, as well as with the increasing storage capacities of modern computers.

When a sufficient amount of VA signal data for the machine or structure in various states has been gathered, the pattern-recognition approach may be applied, provided that some rational approach is adopted for reducing the classification problem for continuous VA signals to the classical pattern-recognition problem for a finite number of variables. One such approach, proposed in [25], uses the so-called Kullback-Leibler measure of information as an index of the 'distance' between any pair of random signals. That is, the degree of closeness between two continuous signals is calculated according to a simple formula in terms of their probability densities*. Moreover, the concept of the nearest-neighbour signal was introduced in [25], which may permit one to decrease the required amount of experimental data. Specifically, for each of the distinctly different states of the machine or component (for example, the normal one and those with certain specific faults), the corresponding reference or baseline VA signal is introduced and stored in the computer; these are regarded as representative of the corresponding states. Any new given VA signal is then classified as representing the same state as the nearest reference signal.

This assumption is a rather crude one (the borderline between cats and dogs is assumed to lie midway between a 'baseline' cat and a 'baseline' dog), as is the assumption of the form of probability densities of the signals,

---

* This, as well as certain other concepts of the theory of random processes, referred to in this Introduction, are defined in Chapter 1

adopted in [25]. Nevertheless, the results of an experimental verification of the approach are rather impressive [25]. It is interesting that the three reference or baseline VA signals in [25] had a rather similar frequency content; at a first glance, the differences between them might be regarded as of secondary importance. However, the algorithm, based on the introduced distance measure, clearly discriminated between these signals. This illustrates one of the advantages of the empirical pattern-recognition approach: it does not rely on the skill and/or experience of the operator and may be completely automated; the operator need not be bothered with detecting and identifying the differences between the measured signals since these are carried out by the computer. On the other hand, though, if the operator himself detects some differences between, say, the frequency content of the measured VA signals in certain frequency ranges, this additional information on the variations of the frequency content of a VA signal may be quite helpful for, say, the localisation of faults. Therefore, more sophisticated algorithms for signal processing may indeed prove to be useful.

In any event, the pattern-recognition approach seems quite promising in situations where there are many parts and components, with good possibilities for gathering a sufficient amount of experimental data, and/or in cases where the requirements of the VA diagnostics are somewhat limited (e.g. detection of faults, which may belong to a limited set of possible faults). For such situations, the advantages of a complete automation of the diagnostics, not requiring a deep insight into the physics of the VA signal generation, may be fully exploited.

Other examples of possible engineering applications of the empirical pattern-recognition approach to VA signals are the discrimination between two different types of water boiling (film-type and bubble-type) within closed vessels (this may be important for the control of heat transfer rates) and the detection of cavitation within the flow through valves. In these applications, the implementation of different regimes of operation, with appropriate noise measurements, seems to be relatively simple. It may also be added that the empirical approach may rely on 'common' statistical analysis as well, such as spectral or correlational techniques, outlined in Chapter 1. For example, the procedure for the detection of cavitation may be based on simple comparisons between the frequency contents of noises in cavitating and noncavitating flows, as obtained from numerous measurements [14].

Another example from the field of nuclear-power technology [32] is now considered, for which the above empirical approaches seem impracticable. An important element of the PWR (Pressurised Water Reactor) is its core barrel, which consists of a vertical cylindrical case (pressure vessel), supported along the periphery of a certain cross-section in its upper part. This core barrel may oscillate during plant operation, mostly because of dynamic disturbances from incoming and outgoing coolant flows. These oscillations are of various types or modes but, from the standpoint of the safety or reliability of the PWR, the monitoring of rigid-body or pendulum-type oscillations of the core barrel is of primary importance, since these oscillations may directly influence the integrity of the barrel's support. Therefore, the problem is to estimate the level of such oscillations by an appropriate processing of the overall vibrational signal of the core barrel. (In fact, even that signal could not be measured directly in [32], and the authors had to rely on ex-core measurements of the neutron flux fluctuation; this fact, however, is not especially important from the standpoint of the diagnostics algorithm, and for simplicity these measured fluctuations may be assumed to be proportional to the dynamic displacements of the core barrel.)

This problem may be classified as that of separation of a useful signal from a noise background; a great variety of such problems have been considered in relation to radar and/or sonar applications. In this case, the 'useful' signal is that produced by the pendulum-type oscillations of the core barrel, whereas all other components of the measured neutron flux signal may be regarded as an extraneous 'noise'. This problem was solved in [32] by estimating the frequency range of the useful signal, with the use of a so-called coherence function, and subsequently filtering the measured signal in this frequency range. As reported in [32], the estimated level of the filtered signal, in which the 'useful' component is dominant, was found to be ten times lower than that of the original one. Therefore, the solution of the diagnostic problem has proved to be really crucial in obtaining a realistic estimate of the reliability of the core-barrel supports.

That example has been presented in this Introduction in order to illustrate the case where the empirical approach does not seem practicable and adequate. The alternative to such an approach is a method of diagnostics which is based, to a certain degree, on a dynamic model of the VA signal generation. This may be a fully quantitative mathematical model, in which case a classical identification problem arises: to formulate a full mathematical description for a given dynamic system by the appropriate processing of its measured responses; by 'full' we mean a description that implies a quantitative estimation of the values of all parameters of the mathematical model. In many cases, though, such a complete quantitative identification may not be necessary for diagnostic purposes, and may also be impracticable. Qualitative and/or semiqualitative identification may then be sufficient. The first of these concepts, introduced in [21], refers, for example, to the classification problem for dynamic systems (based, however, on analyses of the properties of their responses rather than on a direct application of empirical pattern recognition to VA signals). The second concept implies that only a few key quantitative indices for a given dynamic system are sought, such as its stability or reliability margin, whereas a complete quantitative estimation of all the parameters in the mathematical model is once again not required. Incidentally, in the above example of a PWR core barrel, the method of diagnostics in [32] was also based on a qualitative model. Specifically, a feature of pendulum-type oscillations is the synchronous motion of any pair of diametrically opposite points of the barrel. Therefore, to separate the contribution of these motions from the overall measured signal, one may filter this signal in such a way as to suppress all the frequency components outside the range of completely synchronous motions of opposite points. This frequency range was estimated in [32] by using a so-called coherence function, as explained in Chapter 1, according to the condition of a 180°-angle phase shift between signals from two neutron detectors, placed near opposite sides of the core barrel along the same diameter.

This approach to VA diagnostics, which may be called an identificational one, requires, in general, some experience and/or skill from the operator. Moreover, the necessary VA signal-processing procedures may prove to be quite sophisticated and complex. This is especially true if so-called on-line diagnostics is required; that is, if measurements are made of 'natural' VA signals in a normally operating machine or structure, rather than of those in special tests, with artificially applied diagnostic input signals. These natural VA signals are very often quite irregular or random, so that the application of statistical methods for data processing may be required. On the other hand, these sophisticated procedures can, in general, provide much more useful information on the internal state of the machine or structure than can be gained from the purely empirical approach. Moreover, for a newly built machine (structure), this identificational approach seems to be the only possible one, in view of the lack of any reference VA signals.

In many cases, the identificational approach may also provide some additional information, which can be used to improve the machine or structure performance by introducing some appropriate design modifications. This is true especially for cases where a full quantitative identification is made. For example, information on the system's natural frequencies may be used to avoid dangerous resonances or, say, the overlapping of the natural-frequency ranges of the elastic and rigid-body modes of a space vehicle; such overlapping may lead to difficulties with the attitude control of the vehicle.

It seems appropriate now to pay some attention to the terminology in this field. In fact, in several works, classifications of VA diagnostic problems, as well as certain basic concepts, have been introduced [7,11,13,34,53]. In particular, the concept of 'mechanical signature analysis' has become quite common [11]. The meaning of this term seems to be roughly the same as that of the term 'VA diagnostics', as used in this book. Here, VA diagnostics are considered as being either empirical, or 'identificational', as explained above; according to this definition, for example, the estimation of a system's natural frequencies from its response signals also implies its diagnostics. The term 'identificational' refers here to operations with a mathematical model of a machine or structure, or its components; therefore, diagnostics represent one possible field for the application of various identificational techniques — namely to machines and structures.

The classification of possible VA diagnostics problems, as adopted by the present authors, is by no means final, and therefore they have preferred not to present it in, say, the form of a flow chart. It should be regarded rather as a list of possible formulations of the problem and the approaches to their solutions; any possible addition to such a list would be welcome. In any case, the classification is made according to the kind of information that can be gained from a VA signal analysis. This information may be related to certain overall properties of the machine or structure, or to those of its parts or components. For example, since in general any VA signal is not measured exactly at the same place where it is generated, the problem may be to evaluate the properties of the transmission path(s) of the signal(s) within the machine or structure. Another possible problem is to reconstruct the original input signal(s) from measured one(s). With several input excitation sources, the problem of their 'decoupling', or decomposition, may arise; specifically, that of estimating the contribution of each source to the overall measured response signal.

The key to the success of VA diagnostics seems to lie in the proper choice of those features of VA signals that are truly informative of the item under consideration and, moreover, are sufficiently sensitive in this respect. Assume, for example, that the problem is to detect a crack in a rotating shaft from the signal of an accelerometer, fixed to a bearing case. The simplest approach is to rely on the measurements of the intensity level of the overall acceleration signal. However, this crude approach may prove to be insufficiently sensitive for the timely detection of a dangerous situation: the required warning may come too late, when the shaft's vibration level becomes too high. On the other hand, the growth of the acceleration level in certain specific frequency bands, particularly in the vicinity of the rotational frequency and some of its multiples, may be used as a more sensitive indicator of crack growth. Such a criterion for crack detection may provide a sufficiently early warning of imminent rotor failure, while the overall vibration level is still sufficiently low, and thus is far from being dangerous.

A similar example is that of a liquid-propellant rocket engine. Throughout model tests on such engines, noise measurements were made in [47], both for normal operational conditions and for those of imminent combustion instability. The results showed that, while the corresponding variations in the overall noise level were quite moderate, more than tenfold variations in the levels of certain narrowband frequency components could be observed. Thus, while the overall noise level of a liquid-propellant rocket engine does not provide a timely warning of an approaching instability threshold, on using only a slightly more sophisticated processing algorithm for this measured noise signal (including bandpass filtering) one can obtain a more effective estimate (perhaps the simplest one) of the stability margin.

Thus, in this book the emphasis is on this aspect of VA diagnostics: namely on establishing the most informative features of VA signals. Of course, this is by no means the only problem facing the designer of a diagnostic system. Besides a selection of these features for their subsequent use as diagnostic criteria, he also faces many other problems, such as prescribing limiting or threshold values for the selected parameters, corresponding to changes in internal conditions; or integrating all the 'local' diagnostic algorithms into a high-level overall diagnostic system. These topics are considered rather briefly here, as well as those related to a hardware implementation of these algorithms. As for the latter, it may be noted here that, although mostly on-line diagnostic procedures are considered in this book, namely those that may be based on VA measurements in the course of normal operation of the machine or structure, no special requirements for real-time signal processing are considered (such as those that may be needed for automatic control systems).

The contents of the book are arranged as follows. In the first four chapters, the fundamentals of the several various disciplines that are the basis for procedures of mechanical signature analysis or VA diagnostics are briefly outlined: namely the theory of probability and random processes, of signal processing, and of mechanical oscillations and waves. Therefore, these chapters are introductory, and necessarily contain formulae and equations, used mostly to define certain basic concepts. These chapters are included to make the book self-contained; for a more detailed study of these topics, the reader is referred to the specialist literature.

In Chapter 5, certain basic concepts are introduced relating to the dynamic characteristics of linear systems; methods of their evaluation from tests, or in-service measurements, are outlined. These characteristics are used in Chapter 6, where certain quantitative identification problems and methods are considered, such as those of evaluating the transmission paths for vibrational energy and the decomposition of several additive excitation sources. In Chapter 7, so-called semiqualitative identification problems are considered, such as those of estimating the stability margin and identifying parameter variations from measured responses. Chapter 8 is concerned with purely qualitative identification; namely discriminating between various types of dynamic system by an appropriate processing of their on-line response signals. Then, in Chapter 9, several algorithms are presented for the detection of damage, such as cracks, from measured response signals. Finally, Chapter 10 considers diagnostics for rotating machinery.

Chapter 1
Fundamentals of the Theory of Random Processes

Very often, the dynamic loads and disturbances in machines and structures are irregular or random. Therefore, the VA signals in such machines, structural components etc. are also random, and statistical methods based on the theory of probability and random processes are the most appropriate for their analysis. In this chapter, the basic concepts of this theory are presented; they will be widely used in subsequent chapters. This presentation is quite short; for additional details, the interested reader is referred to excellent specialist textbooks, such as [17,46].

The probabilistic approach is used generally for a quantitative description of situations with uncertain outcomes. For example, a single throw of a dice may have six possible outcomes, with equal chances for all of them (provided that the dice and the dicer are both honest). Any specific outcome is therefore a random variable, with possible integer values from one to six. The quantitative measure of the chances for implementation of a certain possible outcome is its probability. It is defined as the relative frequency or ratio of the number of trials with this specific outcome to the total number of trials, when the latter increases indefinitely. By 'trial', we mean here the exact reproduction of the conditions leading to the same set of possible outcomes, with the same chances for each of them in each trial. It is obvious that the total sum of probabilities of all possible outcomes should always be unity and, in general, the probability of any event should be not less than zero and not higher than unity. In the above example of the 'honest' dice, the probability of any outcome of a single throw is easily calculated as being one-sixth, since all six outcomes have identical probabilities. As a more sophisticated example, which may be of some interest to gamblers, the following problem may be considered. Ten balls are picked at random from a group consisting of 16 white and four black balls: find the probability that the selected subgroup of ten balls will contain all four black ones. Or, the original group may consist of 14 white, three red and three blue balls, and it may be of interest to find the probability of finding all three red, as well as all three blue, balls in the selected subgroup of ten balls.

A continuous random variable X, which may assume any real value, is completely described by its so-called cumulative probability function P(x) = Prob (X < x) or by the derivative of the latter, namely a so-called probability density function p(x) = dP(x)/dx. This density defines the probability of the double inequality $x \leq X < x + dx$ for vanishingly small dx. In practice, the estimate of p(x) from a set of observed values of X (outcomes!) is obtained by dividing the whole range of observed values into small finite intervals Δx. Then the relative frequency of the observed values of X within each interval is counted, leading to a histogram, which provides a stepwise-constant approximation to p(x) (Fig. 1.1). (Of course, such an estimate is only

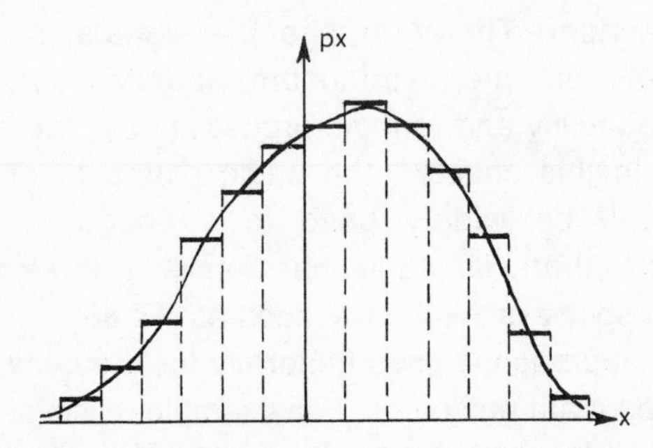

Fig. 1.1 Probability density of a continuous random variable X, together with the corresponding histogram

approximate because of another factor: namely the finite sample of the observed values of X; this point is considered in the next chapter, where certain estimation procedures for random processes are studied.)

As in the case of a discrete random variable, the probability density p(x) of any continuous random variable X is nonnegative everywhere, and moreover

$$\int_{-\infty}^{\infty} p(x)\, dx = 1 \tag{1.1}$$

that is, the area under the curve of p(x) is unity.

Alternatively, a random variable X may be described by a set (infinite) of its so-called moments:

$$m_{Xk} = \langle X^k \rangle = \int_{-\infty}^{\infty} x^k \, p(x) \, dx \quad , \quad k = 1, 2, \ldots \tag{1.2}$$

(here, and in the following, the angular brackets denote probabilistic averaging) or the so-called central moments, i.e. moments of a zero-mean part of X:

$$\mu_{Xk} = \langle (X - m_{X1})^k \rangle = \int_{-\infty}^{\infty} (x - m_{X1})^k \, p(x) \, dx \quad , \quad k = 2, \ldots \tag{1.3}$$

The two lowest-order moments are usually considered to be the most important ones. The first-order moment, m_{X1}, is called a mean or expected value of X, and the second-order central moment $\mu_{X2} = \sigma_{X2}{}^2$ is called a variance of X. Thus, $\sigma_X = \sqrt{\mu_{X2}}$, or the so-called root-mean-square (rms) value of X, is the simplest measure of the scatter of the random variable X around its mean or expected value. Special titles are used also for the third-order central moment μ_3, which is called 'skewness' and characterises the asymmetry of the probability density with respect to its mean value; and for the fourth-order central moment, namely the quantity $æ = (\mu^4 - 3\mu_2{}^2)/\mu_2{}^2$, which is called the 'kurtosis factor'. The skewness is used, for example, for roughness measurement of various machined surfaces; it may show whether the asperities on a given rough surface have flat or peaked summits and thus provide a quantitative index that may be important for estimating wear, fatigue etc.

A pair of random events A and B may be described by their so-called joint probability P(A,B), which may also be expressed as

$$P(A,B) = P(A|B)\,P(B) \tag{1.4}$$

Here, $P(A|B)$ is the so-called conditional probability; that is, the probability of the event A provided that the event B has already occurred. The chances for the event A may be independent of B, in which case $P(A|B)$ is simply P(A) and the joint probability of both events, A and B, equals the product of their individual probabilities; such events are called statistically independent. Similarly, for a pair of continuous random variables X and Y, the joint probability density may be introduced as $p(x,y) = \text{Prob}\,\{x \le X < x + dx\ ,\ y \le Y < y + dy\}$, with vanishingly small dx, dy. From p(x,y), various moments of X and

Y may be calculated, as well as their so-called mixed moments:

$$m_{XkYl} = \langle X^k Y^l \rangle = \int_{-\infty}^{\infty}\int_{-\infty}^{\infty} x^k y^l p(x,y)\, dx\, dy \tag{1.5}$$

or central mixed moments:

$$\begin{aligned} \mu_{XkYl} &= \langle (X - m_{X1})^k (Y - m_{Y1})^l \rangle \\ &= \int_{-\infty}^{\infty}\int_{-\infty}^{\infty} (x - m_{X1})^k (y - m_{Y1})^l p(x,y)\, dx\, dy \end{aligned} \tag{1.5'}$$

The simplest of these, namely m_{X1Y1}, is called the covariance factor. It is convenient to normalise its centred counterpart to the product of the corresponding rms values, since it can be easily shown that the absolute value of $\gamma = \mu_{X1Y1}/(\sigma_X\, \sigma_Y)$ cannot exceed unity. This normalised factor is a suitable single quantitative index of the degree of coupling between two random variables. The extreme cases of $\gamma = 1$ or $\gamma = -1$ correspond to full or ideal coupling, which is possible only in the case of a linear proportionality, or $X = aY$ with $a > 0$ or $a < 0$, respectively. The other extreme case $\gamma = 0$ may correspond to independent random variables X and Y, or, at least, to the case of small statistical coupling.

A function X(t) is called random if its values at any $t = t_i$ are random variables. For the case where the argument t is time, X(t) is usually called a random process. While a single random variable X may be described by its probability density p(x), for a random process X(t) one should use, in general, a joint probability density $p(x_1, x_2, \ldots, x_n; t_1, t_2, \ldots, t_n)$ of the values $X_i = X(t_i)$ of X(t) at time instants $t_1, t_2, \ldots, t_n$, with infinite n. It is clear that such a complete description is impracticable unless a particular model is used that defines a rule for evaluating such an infinite-dimensional probability density in terms of a certain finite set of statistical characteristics of X(t); otherwise, the description of X(t), based on a finite set of its characteristics, will be incomplete. Two such models are of particular importance for statistical dynamics, namely that of the normal or Gaussian random process and that of the Markov process. The first of these models is considered in detail later in this chapter.

The probability density p(x,t) of the value of X(t) at a single time instant may, in fact, be of the same form as that of a random variable X, especially in the case where it does not depend on t. This one-dimensional probability density is one of the basic statistical characteristics of a random process, and its moments may be defined similarly to the case of a random variable, according to eqns. (1.2) and (1.3); in general, these moments may depend on t.

The next element in the hierarchy of probability densities is $p(x_1,x_2,t_1,t_2)$, which relates to the values $X(t_1)$, $X(t_2)$ of $X(t)$ at two different time instants t_1, t_2. Among all the moments of $X(t)$, its mixed second-order central moment is especially important, namely the so-called autocorrelation function of $X(t)$:

$$K_{XX}(t_1,t_2) = \langle [X(t_1) - m_{X1}(t_1)]\,[X(t_2) - m_{X1}(t_2)] \rangle$$

$$= \int_{-\infty}^{\infty}\int_{-\infty}^{\infty} [x_1 - m_{X1}(t_1)]\,[x_2 - m_{X1}(t_2)]\, p(x_1,x_2,t_1,t_2)\, dx_1\, dx_2 \qquad (1.6)$$

This function is a basic characteristic of statistical coupling between the values of a random process at two different time instants. Its normalised version $r_{XX}(t_1,t_2) = K_{XX}(t_1,t_2)/[\sigma_X(t_1)\,\sigma_X(t_2)]$ satisfies the inequalities $-1 \le r \le 1$, similarly to the case of two random variables.

A very important concept is that of a stationary random process. In general, stationarity means the independence of some properties of a random process $X(t)$ on a shift of time origin. Whenever such independence is implied, for all statistical characteristics of $X(t)$, the process is called strictly stationary. If such a requirement concerns only the most important first- and second-order moments of $X(t)$, the process is called stationary in a wide sense. In the latter case, its expected value m_{X1} is independent of t, whereas its autocorrelation function $K_{XX}(\tau)$ depends only on the time shift $\tau = t_2 - t_1$ rather than on both t_1 and t_2. The property of stationarity is very closely related to that of a so-called ergodicity. A process $X(t)$ is called ergodic if its mean value, as defined by a probabilistic, or so-called ensemble, averaging, according to eqn. (1.2), coincides with the result of a time averaging; namely:

$$\langle X(t) \rangle = \lim_{T\to\infty} \frac{1}{T}\int_0^T X(t)\, dt \qquad (1.7)$$

This implies, roughly speaking, that, if a sufficiently long sample of $X(t)$ is considered (theoretically it should be of an infinite length), then essentially all features of $X(t)$ will be represented in this sample. In this case, it is possible to use for averaging, according to eqn. (1.7), the values of X as selected from this single sample rather than to use ensemble averaging according to eqn. (1.2); the latter involves picking simultaneous values of X(t), at a single time instant, from numerous samples. It is obvious that this equivalence is of the utmost importance for the processing of measured random VA signals.

The procedures for testing whether the measured random signal is stationary, or not, are, roughly speaking, mostly based on dividing the recorded signal, or available sample, of the process into several parts and estimating its various statistical characteristics independently for each of these different parts of the sample. For a stationary process, these estimates, as obtained for different parts of the sample, should be more or less identical. By 'more or less', we refer to the inevitable statistical scatter or uncertainty of the estimates, as obtained from finite-length samples. This is indeed an inherent difficulty for checking the stationarity property; one may be fairly sure of the results, but not with 100% certainty.

For limited available experimental data, some hypotheses are usually introduced concerning the stationarity properties of a random process. Very often such a hypothesis may be accepted in the case of constant 'macroconditions' for the generation of the given random process, such as for turbulent pressure oscillations in a steady fluid flow. For example, the noise of a jet aircraft, as heard by passengers, may be regarded as a stationary random process if the engine operates in a constant regime, particularly with a fixed rotation speed of the compressor shaft*; for example, during engine warming-up or free taxiing of the aircraft from a parking place to the runway. On the other hand, when the engine rotation speed, and thus the jet velocity, is increased, say, on the runway just before takeoff, definite changes in audible noise can be clearly heard. Specifically, this noise increases and also has a higher tone (its dominant frequencies are increased). Another example is that of a road with a wavy, or rough, surface. The elevation of this surface may be generally regarded as a random function of the longitudinal coordinate along the track of the corresponding point of the road surface. This function may be regarded as being stationary, provided that the conditions for the formation of waviness were uniform. Consider now a vehicle, travelling along such a rough road. It may be shown that, in this case of spatial stationarity, the excitation of the vehicle's wheels due to irregularities in the track profile will be stationary random processes, provided that the vehicle travels with a constant speed. In the case of a variable speed, say, of an aircraft during acceleration along a runway just before takeoff, the excitation and therefore the vehicle's vibrational response are definitely not stationary.

---

* With a certain reservation to be considered shortly

As for VA signals in machines and structures, they may sometimes be nonstationary, first of all, because of impact-type loading, say, during an earthquake. A certain rather specific source of nonstationarity in the course of steady operation of a machine or structure is a periodic excitation. It may lead to 'periodically nonstationary' VA signals, their statistical characteristics being periodic functions of time. In the above example of a jet-engine noise, the acoustic signal may, in general, contain a periodic component of the passage frequency of the compressor rotor blades, leading to a so-called pure tone. Certain procedures for the processing of periodically nonstationary random processes are considered in Chapters 2 and 9.

The basic statistical characteristics of a stationary random process X(t) are now considered in further detail. Its probability density p(x) generally does not depend on time. A great variety of VA signals have a bell-shaped p(x), as shown in Fig. 1.2(a). Moreover, in many cases, these signals may be

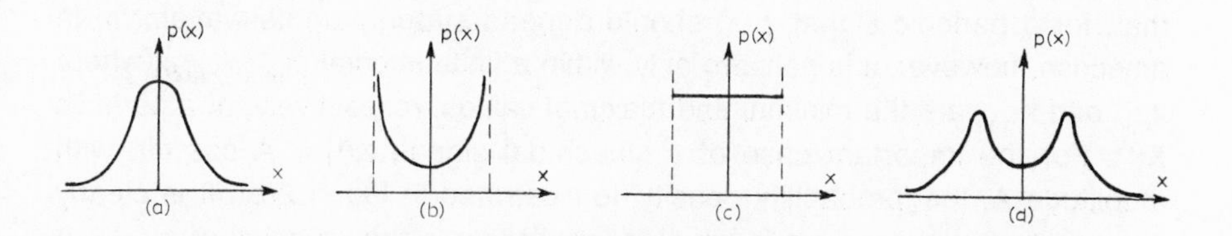

Fig. 1.2 Various shapes of probability densities of zero-mean VA signals: (a) bell-shaped for a purely random signal, (b) singular for a smooth periodic signal, (c) uniform for a sawtooth signal, (d) smooth bimodal for a mixture of periodic and random signals

considered as normal or Gaussian, i.e.

$$p(x) = (\sigma_X \sqrt{2\pi})^{-1} \exp\left[-\frac{(x - m_X)^2}{2\sigma_X^2}\right] \tag{1.8}$$

where m_X and σ_X are the mean and rms values of X(t), respectively. Gaussian processes are of the utmost importance for statistical dynamics, as shown in Chapter 3; most of their basic properties are the consequences of the so-called Central Limit Theorem [17,46]. Thus, the probability density of a sum of independent random variables (or stationary processes) approaches the normal or Gaussian one if the number of components increases indefinitely regardless of the individual probability densities of those components.

The normal distribution p(x) is symmetric with respect to the mean value of X, i.e. $\mu_3 = 0$. Moreover, its kurtosis is zero; in fact, the kurtosis factor $æ_4$ is just the index of the degree of closeness of a given probability density to the normal one. Such an index is used sometimes for detecting growing faults in rolling element bearings from their measured VA signals [45].

The bell-shaped probability, as illustrated in Fig. 1.2(a), is typical of purely random, original VA signals. A rather different case is that of a periodic signal, with its waveform exactly reproduced after a certain time interval, or period T. Of course, since such a signal is not random in time, the concept of a probability density, rigorously speaking, is not appropriate in this case. However, it may be of use, nevertheless, for VA diagnostics, as will be shown in more detail in subsequent chapters. Indeed, the cumulative 'probability' function P(x) may be defined for a periodic X(t) with period T as the relative stay time of X(t) under a certain fixed level x: $P(x) = \theta/T$, where θ is the total length of those time intervals, within $0 \leq t < T$, where $X(t) < x$. 'Probability density' p(x) may be defined then as the derivative $p(x) = dP/dx$. It is obvious that, for a periodic signal, p(x) should depend strongly on its waveform. In any case, however, it is nonzero only within a finite interval (x_{min}, x_{max}), where x_{min} and x_{max} are the minimal and maximal values, respectively, of a periodic X(t). For the important case of a sinusoidal signal, $X(t) = A \cos \omega t$, with amplitude A, the 'probability density' is illustrated in Fig. 1.2(b). It is clearly seen that p(x) in this case is growing indefinitely in the vicinities of $x_{max} = A$ and $x_{min} = -A$. The reason is the vanishingly small rate of the time derivative of such a smooth signal at these extreme points. For example, a signal of the 'sawtooth' type waveform with sharp peaks does not have such singularities of p(x); in fact, it has a uniform 'probability density'; i.e. p(x) is constant everywhere within $x_{min} \leq x \leq x_{max}$, as shown in Fig. 1.2(c). Thus, this sawtooth signal is a convenient test signal for checking algorithms and/or electronic devices for probability density measurements.

It is clear that a random or noise-type signal, added to a sinusoidal one, should lead to a smoothing of the singularities of p(x) at $x = \pm A$ (in this case, the maximal and minimal values of X(t) may not exist at all if an infinite sample is considered). Therefore, for the case of a relatively small random component, p(x) should have an appearance as shown in Fig. 1.2(d). Such a typical shape of probability density for a mixture of sinusoidal and random signals is used sometimes for the detection of a periodic component in a measured VA signal. This probability is definitely bimodal; that is, it has two maximum points, contrary to the unimodal bell-shaped p(x) of Fig. 1.2(a). However, in the case of a relatively high random component compared with the sinusoidal one, the probability density of their mixture may become unimodal, as in Fig. 1.2(a).

We illustrate here now two more examples of one-dimensional probability densities, which are of importance for VA diagnostics. Consider two independent stationary normal random processes X(t) and Y(t) with zero mean values and identical rms values $\sigma_x = \sigma_y = \sigma$. Denote their sum of squares as $V(t) = X^2(t) + Y^2(t)$, and let also $A(t) = \sqrt{V(t)}$. It may be noted that these processes A(t) and V(t) are very important for so-called narrowband random processes, as is explained later on in this chapter, and are also widely used for VA diagnostics. The probability densities of A(t) and V(t) are governed by the following formulae, as derived from eqn. (1.8) with the use of the well known probability transformation laws [21,46]:

$$p(A) = \left(\frac{A}{\sigma^2}\right) \exp\left(\frac{-A^2}{2\sigma^2}\right) \tag{1.9}$$

$$w(V) = \left(\frac{1}{2\sigma^2}\right) \exp\left(\frac{-V}{2\sigma^2}\right) \tag{1.10}$$

These probability densities, as defined within positive semiaxes, of the strictly positive random functions A(t) and V(t), are illustrated in Figs. 1.3(a) and 1.3(b), respectively. While the former has a mode that is a maximum at the

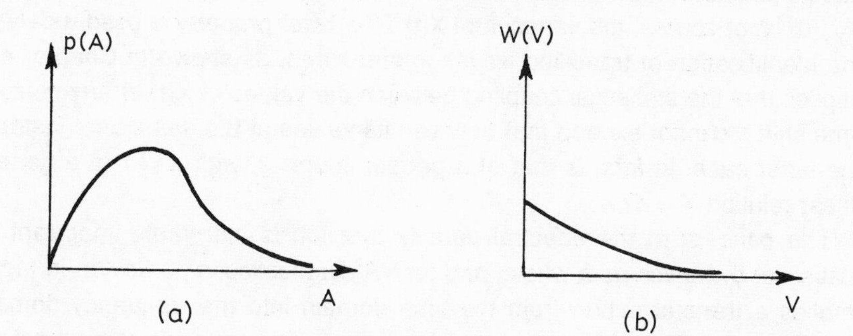

Fig. 1.3 Probability densities of the amplitude of a Gaussian random process (a) and of its square $V = A^2$ (b)

most probable value of A(t), the second one monotonically decreases at every $V \geq 0$. It is shown later that the lack of such a monotonous decrease of w(V) for a measured VA signal can have important implications for the type of dynamic system with such a response signal.

The statistical characteristics of the values of a zero-mean stationary random process X(t) at two different time instants, $X_1 = X(t_1)$, $X_2 = X(t_2)$, are now considered. The most complete of these is the two-dimensional probability density $p(x_1,x_2,\tau)$ of X_1 and X_2, which for a strictly stationary X(t) depends only on the time shift $\tau = t_2 - t_1$ rather than on both t_1 and t_2. However, for VA diagnostic purposes, the most widely used characteristic is either the second-order moment; that is the autocorrelation function of X(t), defined by

$$K_{XX}(\tau) = \langle X(t_1)\, X(t_2)\rangle = \int_{-\infty}^{\infty}\int_{-\infty}^{\infty} x_1\, x_2\, p(x_1,x_2,\tau)\, dx_1\, dx_2 \tag{1.11}$$

or its so-called Fourier transform, or spectral density function,

$$\Phi_{XX}(\omega) = \frac{1}{2\pi}\int_{-\infty}^{\infty} K_{XX}(\tau)\, e^{-i\omega\tau}\, d\tau = \frac{1}{\pi}\int_{0}^{\infty} K_{XX}(\tau)\cos\omega\tau\, d\tau \tag{1.12}$$

The autocorrelation function depends only on the time shift $\tau = t_1 - t_2$, both for strictly stationary random processes and for processes that are stationary in the wide sense. Moreover, it is an even function; that is, $K_{XX}(-\tau) = K_{XX}(\tau)$, and has its absolute maximum at zero time shift, $K_{XX}(\tau) \le K_{XX}(0)$ for any τ, where $K_{XX}(0)$ is, of course, the variance of X(t). The latter property is used widely for the identification of travelling waves in structures, as shown in Chapter 4. It implies that the statistical coupling between the values of X(t) at any nonzero time shift τ cannot exceed that between its values at the same time instants; the latter case, in fact, is that of a perfect coupling, with a = 1 in a general linear relation X = aY.

The concept of the spectral density function is extremely important for statistical dynamics as a whole, and for VA diagnostics in particular. In fact, it implies a transformation from the time domain into the frequency domain. This is important in emphasising various specific frequency components of structural responses to random excitation. Thus, it defines, in fact, the frequency distribution of the energy of a given process; for a stationary random X(t), this distribution is continuous rather than discrete, as for a periodic or 'quasiperiodic' signal. In fact, the spectral density defined in eqn. (1.12) may be obtained in another way; namely by considering X(t) as a sum of sinusoidal functions $a_i \sin \omega_i t$ with independent random (constant) amplitudes a_i and an indefinitely increasing number of component frequencies. This representation is widely used for the direct estimation of the spectral density from a finite length record of a measured random signal.

It can be shown that $\Phi_{XX}(\omega)$ is nonnegative everywhere (as an energy distribution function should be), and

$$K_{XX}(\tau) = \int_{-\infty}^{\infty} \Phi_{XX}(\omega)\, e^{i\omega\tau}\, d\omega = 2\int_{0}^{\infty} \Phi_{XX}(\omega) \cos \omega\tau \, d\omega \qquad (1.13)$$

Since $K_{XX}(0)$ is the variance of $X(t)$, the special case of eqn. (1.13) for $\tau = 0$ has a clear physical interpretation: the total energy of a process is, in fact, a simple sum of the energies of its various frequency components.

Since $K_{XX}(\tau)$ and $\Phi_{XX}(\omega)$ are interrelated according to eqns. (1.12) and (1.13), they may in principle be regarded as equivalent characteristics of a random process. For certain specific applications, one of these functions may be more suitable than another. For example, in resonant vibrations, certain specific frequency components of the response may be emphasised; in this case, the frequency-domain approach, through the use of spectral density, may be more convenient. On the other hand, the response of a large structure may sometimes be mostly of a travelling-wave type, with some reproduction of the responses at various points being possible, after certain time shifts; for such cases, the correlational approach may be preferable.

Figure 1.4 illustrates the qualitative nature of $K(\tau)$, and corresponding

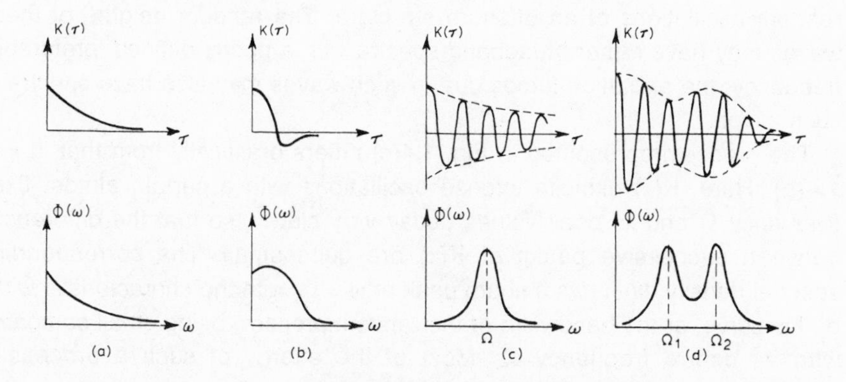

Fig. 1.4 Examples of autocorrelation functions $K(\tau)$ of stationary random processes (upper curves) and corresponding spectral densities $\Phi(\omega)$ (lower curves)

$\Phi(\omega)$, for certain cases. In Fig. 1.4(a), $K(\tau)$ is positive everywhere, whereas $\Phi(\omega)$ is monotonically decreasing for every $\omega \geq 0$. Such spectra are typical, for example, of turbulent oscillations in fluid flows, at least in cases of unseparated and unconfined flow — in the absence of any vortex shedding and acoustical resonant phenomena. The higher the decay rate of $K(\tau)$, the lower that of the spectral density, with the energy of the process being distributed fairly uniformly within a rather broad frequency range. In the limit of a vanishingly small characteristic time of the decay of $K(\tau)$, that is of the so-called correlation time of the process, $\Phi(\omega)$ becomes constant everywhere, even at infinite frequencies. Such a process is called 'white noise', by analogy with optics, where a white light may be considered as a uniform mixture of light rays of various colours. Such a process may appear somewhat unrealistic because of its infinite energy (the spectral density of any real-life process should decay and vanish at sufficiently high frequencies). However, as will be seen in Chapter 3, such a model of a random process may nevertheless be useful and is, in fact, widely used in statistical dynamics.

In Fig. 1.4(b), another type of stationary random process is represented in terms of its autocorrelation function $K(\tau)$ and spectral density $\Phi(\omega)$. This process has a certain 'preferred' frequency component, corresponding to the peak of $\Phi(\omega)$, while $K(\tau)$ may in general (but not necessarily) be negative at certain values of τ. This peak in Fig. 1.4(b) is not very sharp, so that the characteristic width, or the bandwidth, of the process may be comparable with the value of the 'preferred' frequency Ω itself, at which $\Phi(\omega)$ has its maximum. Such a random process is called broadband. An example of such a process may be found in offshore engineering, where ocean waves excite random oscillations of an offshore structure. The random heights of these waves may have rather broadband spectra with a poorly defined 'preferable' frequency; the excitation forces due to such waves may also have spectra of such a type.

The process represented in Fig. 1.4(c) differs drastically from that in Fig. 1.4(a). Here, $K(\tau)$ exhibits intense oscillations with a certain almost fixed frequency Ω and its peak values decay very slowly, so that the differences between successive peaks of $K(\tau)$ are quite small. The corresponding spectral density $\Phi(\omega)$ has a sharp peak at $\omega = \Omega$, with the characteristic width of this curve, or the bandwidth of the random process, being small compared with its centre frequency Ω. Most of the energy of such a process is concentrated in the vicinity of this frequency. This process is called narrowband.

It may be expected that a narrowband process will have the appearance of a sinusoid with a slowly (randomly) varying amplitude and phase; that is, its peak values and periods should have only small variations between consecutive cycles. In Fig. 1.5, a sample of a computer-generated narrowband

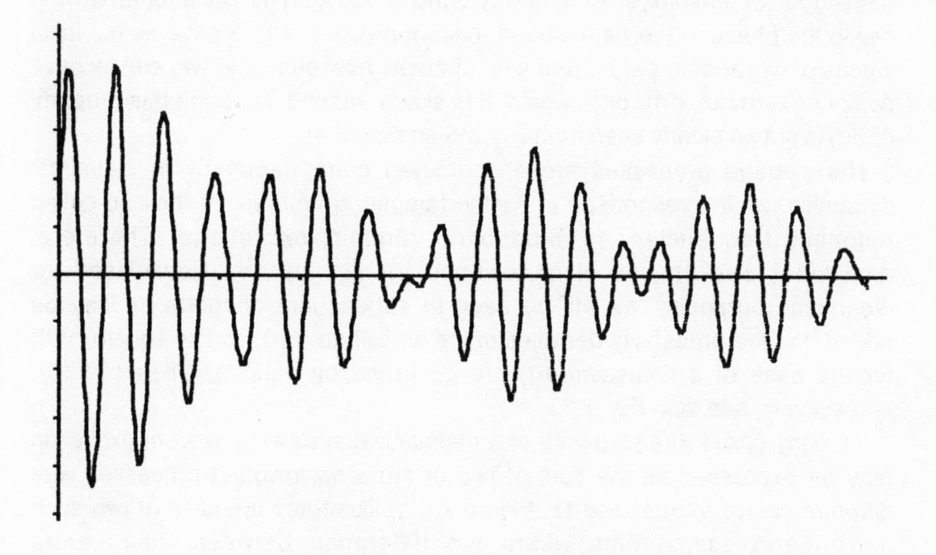

Fig. 1.5 Sample of a computer-drawn narrowband random process

random process x(t) is presented, which illustrates this property. Mathematically, it may be expressed by representing x(t) and its time derivative $\dot{x}(t)$ as:

$$x(t) = A(t) \sin [\Omega t + \varphi(t)]$$

$$\dot{x}(t) = \Omega A(t) \cos [\Omega t + \varphi(t)] \tag{1.14}$$

So that

$$A(t) = \left[x^2(t) + \frac{\dot{x}^2(t)}{\Omega^2} \right]^{1/2}$$

$$\varphi(t) = \arctan \left[\frac{\Omega\, x(t)}{\dot{x}(t)} \right] + \Omega t \tag{1.15}$$

The first of these new random processes, A(t), defines the peaks of x(t) and is indeed slowly varying, for a narrowband x(t) (see Fig. 1.5). It is called the amplitude, or envelope, of a narrowband process. The other one, $\varphi(t)$, is called the phase of the narrowband random process, x(t). It governs the local values of its random period (the time intervals between, say, two consecutive peaks or zeros of x(t)); once again, it is slowly varying, although this property of $\varphi(t)$ is not so clearly seen from Fig. 1.5 as that of A(t).

Narrowband processes are encountered quite frequently in statistical dynamics, as the responses of lightly damped structures, at their so-called resonant frequencies, to broadband random excitations. Therefore, amplitudes and phases of narrowband VA signals are widely used for diagnostic purposes, as will be seen in subsequent chapters. It may be added that the probability densities of the amplitude, A(t), and its square, V(t), for the case of a Gaussian x(t) are governed by eqns. (1.9) and (1.10), respectively; see also Fig. 1.3.

In many cases, the response of a mechanical system to random excitation may be expressed as the sum of two or more narrowband processes with different centre frequencies Ω_i. Figure 1.4(d) illustrates the case of two such narrowband components where the difference between their centre frequencies $\Omega_2 - \Omega_1$, or so-called detuning, is of the same order as their bandwidths. The weighted sum of such processes has a spectral density with a double peak, as long as the detuning is not too small for these peaks to merge into a single one, whereas its correlation function exhibits the so-called 'beat' phenomenon; that is, it has an envelope (a continuous curve drawn through all the consecutive peaks), which does not decay monotonically, as in Fig. 1.4(c), but rather may oscillate with a small frequency of the order $\Omega_2 - \Omega_1$. With decreasing detuning $\Omega_2 - \Omega_1$, the two spectral peaks of x(t) become closer, and may ultimately merge; in this case, the direct detection or resolution of such a frequency split, from a single measured VA signal, may be difficult, or even impossible.

In many empirical diagnostic procedures, which do not rely on a dynamic model of VA signal generation, it is the spectral densities of such signals that are used to discriminate between the various states of the given system. Such procedures are usually based on numerous spectral-density measurements for the signals of the system in cases where its state is known; then the differences in the measured spectra for different states are sought. Many examples of such an approach can be found in hydrodynamics; thus, in [14] it was adopted to discriminate between erosive and nonerosive cavitating liquid jets, using noise measurements.

The concepts of correlation functions and spectral densities may be easily extended to the case of a pair of random processes X(t) and Y(t). Namely, the mixed second-order central moment

$$K_{XY}(\tau) = \langle [X(t_1) - m_{X1}]\,[Y(t_2) - m_{Y1}] \rangle$$

$$= \int_{-\infty}^{\infty}\int_{-\infty}^{\infty} (x_1 - m_{X1})\,(y_2 - m_{Y1})\, p(x_1, y_2, \tau)\, dx_1\, dy_2 \qquad (1.16)$$

is called the crosscorrelation function of the stationary processes X(t), Y(t). The dependence of their joint probability density p(x,y) and crosscorrelation function K_{XY} only on the time shift $\tau = t_1 - t_2$ rather than on both t_1 and t_2 implies that these processes are not only stationary but also 'stationary coupled' (strictly and in a broad sense, respectively). The Fourier transform of $K_{XY}(\tau)$ is called the cross-spectral density of the pair of stationary and stationary-coupled processes X(t) and Y(t):

$$\Phi_{XY}(\omega) = \frac{1}{2\pi}\int_{-\infty}^{\infty} K_{XY}(\tau)\, e^{-i\omega\tau}\, d\tau$$

$$K_{XY}(\tau) = \int_{-\infty}^{\infty} \Phi_{XY}(\omega)\, e^{i\omega\tau}\, d\omega \qquad (1.17)$$

The crosscorrelation function of two random processes is not, in general, even (contrary to their autocorrelation functions), and therefore the corresponding cross-spectral density may be a complex function; in Chapter 5, we consider the physical interpretation of both its real and imaginary parts, for certain cases that arise in statistical dynamics.

An extremely important concept is that of the coherence function $\gamma^2(\omega)$; this is defined as the normalised squared cross-spectral density, namely:

$$\gamma^2_{XY}(\omega) = \frac{|\Phi^2_{XY}(\omega)|}{\Phi_{XX}(\omega)\,\Phi_{YY}(\omega)} \qquad (1.18)$$

This function provides a measure of the statistical coupling between those components of random processes, X(t) and Y(t), that have the same frequency ω. Of course, the values of this function are bounded everywhere in the same way as those of a squared crosscorrelation factor of two random variables: namely $0 \le \gamma_{XY}{}^2(\omega) \le 1$ for every ω. High coherence between two VA signals (such that $\gamma^2(\omega)$ is close to unity) implies that they are strongly coupled within the corresponding frequency range. Lack of coherence (small

$\gamma^2(\omega)$) may often be considered as the result of extraneous noises being added to the measured VA signals. Thus, it was the coherence function that was used in [32] to extract the pendulum-type oscillations of the PWR barrel from the measured overall neutron-flux-fluctuations signal (this problem was described in the Introduction). It may be expected that pendulum-type oscillations of the barrel in a fixed plane should induce ideally coupled signals at two diametrically opposite points with respect to the barrel. Therefore, the procedure was to estimate $\gamma^2(\omega)$ for various pairs of signals, from such points around the circumference, and then to estimate the vibration signal levels only within frequency ranges with values of $\gamma^2(\omega)$ close to unity. The resulting estimates were found to be about ten times less than those obtained directly from the overall measured signals, which contained, therefore, strong extraneous components in certain frequency ranges. By the way, this study showed that the barrel's oscillations were two-dimensional, rather than one-dimensional (in a fixed plane) [32]. It may be noted that the coherence-function concept, and its various versions for the case of more than two random processes, form the basis of the solution to an important problem of VA diagnostics; namely that of 'excitation sources decomposition', considered in Chapter 6.

The above list of various statistical characteristics, which are generally used for VA diagnostics, is by no means complete. They may be regarded simply as the basic ones, which are generally estimated according to general-purpose computer programs of VA signal processing. Many other characteristics are also used for VA diagnostic purposes. For example, third-order moments can be calculated for values of a random process at three different time instants; from these moment, the so-called bi-spectra of a random process may be obtained. Furthermore, certain nonlinear transformations of these basic characteristics may be used also for VA diagnostics, such as those forming the basis for 'cepstral analysis'. These, as well as certain other more sophisticated characteristics of random processes, are considered in subsequent chapters.

Concluding this introductory chapter on the theory of probability and random processes, it should be stressed once again that the presentation here is rather brief, and many topics, not directly related to VA diagnostics, have not been considered at all. One of these topics, however, which is of extreme importance in mechanical engineering, should be mentioned here. This, the theory of probability and random processes, is used widely for reliability predictions for machines and structural components. Such predictions are often based on rather sophisticated characteristics of the random processes involved. These characteristics are found to be different for different types of failures; namely for those due to a single crossing of some 'threshold' or limiting value by the random process (for example, due to brittle fracture of material) and for those due to slow damage accumulation (such as fatigue failure or wear). In any case, however, the reliability is governed mainly by the 'tails' of the probability densities of the random processes involved, or by their higher-order moments. These topics, which are related to the 'theory of excursions' of random processes, are discussed in some detail in [10,17,19,21,46]; they may be of importance for VA diagnostics where the aim is to obtain current life predictions for a machine or structure from its measured VA signal.

Chapter 2
Fundamentals of Signal Processing

This chapter is of a preliminary nature, as was the previous one, and it is also (necessarily) more mathematical than the bulk of the book. Indeed, it is difficult to present the basic concepts and algorithms of signal processing without introducing their mathematical definitions. In subsequent chapters, however, these concepts will be dealt with freely, often without using these mathematical definitions, by appealing only to their physical meaning.

To begin with, a certain rough classification of the signals that may need to be processed seems relevant. First of all, a VA signal may be classified as being either deterministic or random, although this classification should sometimes be regarded as somewhat provisional. Indeed, in some algorithms, the signal is treated simply as a given function of time, irrespective of its nature (deterministic or random). On the other hand, certain statistical processing algorithms are often applied to even seemingly deterministic signals, in order to filter out possible noises and/or errors.

Deterministic VA signals may be classified as being either periodic (i.e. repeating themselves with a certain fixed period) or nonperiodic. The former are found, for example, in rotating machinery. The most common nonperiodic VA signals are the decaying transients excited by impacts or impulse-type loadings. Sometimes, a nonperiodic deterministic signal may be generated as the sum of two (or more) periodic signals from different excitation sources with incommensurable periods; such processes are called almost periodic. As for random VA signals, they may be classified as being either stationary or nonstationary, as explained in the previous chapter.

The most commonly used operation in VA signal processing is, perhaps, Fourier transformation (FT). It maps any signal into the frequency domain (from the time domain), so that this signal is represented as a superposition of its sine and cosine components; as already mentioned in the previous chapter, this approach may be of great use since the various frequency components of a VA signal may be definitely related to certain specific features of the internal state of a machine or structure. In the special case of a periodic deterministic function x(t), which repeats itself exactly with a period T, its FT, X(f), is defined as:

$$X(nf) = \frac{1}{T} \int_{-T/2}^{T/2} x(t) \exp(-2\pi inft)\, dt \quad (2.1)$$

whereas the signal x(t) may be represented by the following sum, or Fourier series:

$$x(t) = \frac{a_0}{2} + \sum_{n=1}^{\infty} (a_n \cos 2\pi fnt + b_n \sin 2\pi fnt) \quad (2.2)$$

where a_n and b_n are proportional to the real and imaginary part of X(nf), namely

$$a_n = \frac{2}{T} \int_{-T/2}^{T/2} x(t) \cos 2\pi fnt\, dt ,$$

$$b_n = \frac{2}{T} \int_{-T/2}^{T/2} x(t) \sin 2\pi fnt\, dt \quad (2.3)$$

Here, f = 1/T is the circular frequency, which is usually measured in cycles per second or Hertz (Hz); also, the angular frequency $\omega = 2\pi f$ is commonly used and is measured in radians per second. It is seen from eqn. (2.2) that the FT of a strictly periodic signal may have nonzero values only at the frequency f = 1/T of this signal and its integer multiples, or so-called harmonics nf, n = 1, 2, ... It is said, therefore, that a periodic signal has a discrete spectrum with step f in the frequency domain. On the other hand, if the period T is increased indefinitely, this step $f = \omega/2\pi$ vanishes. This limiting case corresponds to a nonperiodic signal with a continuous spectrum X(f) (in general, a complex function), as defined according to the general integral formulae for Fourier transforms [4,35]:

$$x(t) = \int_{-\infty}^{\infty} X(f)\, e^{2\pi ift}\, df = \frac{1}{2\pi} \int_{-\infty}^{\infty} X(\omega)\, e^{i\omega t}\, d\omega \quad (2.4)$$

$$X(f) = \int_{-\infty}^{\infty} x(t)\, e^{-2\pi i f t}\, dt \tag{2.5}$$

The integrals in eqns. (2.4) and (2.5) may sometimes be regarded in a generalised sense. For example, a cosine signal $x(t) = \cos \omega_0 t$ has, according to eqn. (2.5), an FT described by Dirac delta-functions $X(\omega) = \frac{1}{2}[\delta(\omega + \omega_0) + \delta(\omega - \omega_0)]$. The delta-function $\delta(u)$, by definition, is zero everywhere except at $u = 0$, whereas $\delta(0) = \infty$ and

$$\int_{-\infty}^{\infty} \delta(u)\, du = 1 \quad .$$

Therefore, the above formula for the spectrum of a cosine signal simply means that this signal contains only a single harmonic component with frequency ω_0 (and $-\omega_0$ as well, since the general formulae for an FT consider both positive and negative frequencies).

Any signal, $x(t)$, and its FT, $X(f)$, satisfy the important equality

$$\int_{-\infty}^{\infty} x^2(t)\, dt = \int_{-\infty}^{\infty} |X(f)|^2\, df \tag{2.6}$$

which implies that the total energy of any signal — by definition the integral of the squared signal over the whole domain where the signal does not vanish — should not depend on the way the signal is represented (in the time or the frequency domain).

The convenience of the FT stems from the fact that both sine and cosine functions do not change their shape after linear transformations, such as differentiation and integration. Moreover, the response of a linear dynamic system with lumped parameters to any frequency component of an input signal may be considered independently of the other components. This topic will be considered in more detail in subsequent chapters, with full definitions of all the concepts involved; in this chapter, we restrict ourselves to a specific class of linear systems, namely filters.

In general, a filter may be defined as a dynamic system, which provides a certain desirable transformation of a given signal (or class of signals), such that it is attenuated and/or magnified in various specific ranges of the time or frequency domain. (These attenuation and/or magnification properties may also be dependent on the level of the signal, but such nonlinear filtering is not considered here; an example of nonlinear filtering, the 'cepstral analysis', is described in Chapter 6.) For VA diagnostic purposes, selective filters in the frequency domain are mostly used. Such filters are classified as high-pass, low-pass and bandpass ones. A bandpass filter suppresses all the frequency components of an input signal outside a given frequency range, say Ω_1, Ω_2, where Ω_1 and Ω_2 are the lower and higher cutoff frequencies, respectively. The basic parameters of a bandpass filter are its centre frequency $\Omega_0 = (\Omega_1 + \Omega_2)/2$ and its bandwidth $\Delta = \Omega_2 - \Omega_1$. Such a filter may be used for a more detailed analysis of the signal in the vicinity of a certain frequency Ω (a sort of a 'magnifying glass' in the frequency domain). In this case, the frequency Ω should lie within the range (Ω_1,Ω_2), preferably close to the centre frequency Ω_0, whereas the bandwidth Δ should be chosen in such a way as to preserve all those frequency components in the vicinity of Ω that may be of interest.

It may be added that, in such filters of the common type, the usual operations of differentiation and integration are used very often. The filtering may be applied directly to a 'raw' electrical signal as obtained, say, from a vibration sensor or a microphone. In such cases, commercial electronic filters are used. Most of these have standard values of relative bandwidth which correspond to one octave ($\Omega_2/\Omega_1 = 2$) or one-third of an octave ($\Omega_2/\Omega_1) = \sqrt[3]{2} = 1.26$). Such electronic filters, in most cases, do not provide a perfect cutoff of the signal outside the desired range, and one should take into account their real amplitude and phase characteristics [28]. On the other hand, such electronic implementation of filtering is not needed for signals that are already converted into a digital form, since, in this case, all the necessary operations may be performed by a digital computer; an extensive treatment of digital filtering may be found in [27].

Both high-pass and low-pass filters may be regarded, in fact, as special cases of a bandpass filter with infinite upper cutoff frequency, and zero lower cutoff frequency, respectively. The former is used for suppressing the low-frequency components of the signal. For example, in systems for measuring the topography of rough surfaces of machine components, the original profilometer signal is usually high-pass filtered in order to suppress low-frequency components, which are related to machining errors ('waviness') rather than to 'true' surface roughness (if the latter is of principal interest). (In the above example, these two types of surface irregularity — namely waviness and roughness — are defined originally in terms of their wavelengths λ, which are higher for the former; the frequency of a profilometer signal, however, is proportional to the spatial frequency $2\pi/\lambda$ of the surface irregularities with a proportionality factor governed by the scanning speed of the profilometer head.)

Low-pass filters are used widely, for example, to minimise sampling errors, which may arise when the original continuous analogue signal is converted into a digital form for subsequent processing on a digital computer. This operation, as performed by a special device, namely an analogue-to-digital converter, or ADC, is used in most VA diagnostic systems. Indeed, usually only relatively simple diagnostic algorithms may be implemented by purely analogue electronic devices, whereas, for more or less sophisticated algorithms of VA signal processing, a digital computer may be required.

An ADC provides 'sampling' of a continuous signal $x(t)$; that is, it transforms the latter into a sequence of its selected values $x_i = x(t_i)$. This sequence is formed usually with a constant sampling rate and then introduced into a digital computer. It is important to make a proper choice of this constant sampling rate, or the constant time interval or step $h = t_{i+1} - t_i$, between each pair of consecutive instants t_i, so that the sampled sequence, or so-called time series, x_i, adequately reproduces all the essential features of the original continuous signal $x(t)$. This choice is based generally on the well known Kotelnikov theorem [29], which may be stated as follows.

A function $x(t)$ with a maximal or upper bound frequency f_o in its spectrum (Fourier transform) may be uniquely reconstructed from its sample values $x_i = x(t_i)$ at equispaced intervals $h = t_{i+1} - t_i$, provided that the sampling rate $f_s = 1/h$ is higher than $2f_o$.

The physical meaning of this statement is rather simple: it implies, in fact, that in the conditions of the theorem a smooth interpolation of the signal between its sample values x_i is possible since $x(t)$ does not contain any components with circular frequencies higher than $f_s/2$. The upper bound frequency f_o is also known in Western literature as the Nyquist frequency, and the minimal admissible sampling rate, $2f_o$, as the Nyquist rate.

If the sampling in the ADC is performed at a lower rate than $2f_o$, an error may be introduced in the resulting time series because of the so-called aliasing phenomenon [22,28]. The point here is that sampling of $x(t)$ is formally equivalent to multiplication of $x(t)$ by a sequence of equispaced delta-function-type impulses $\delta(t - t_i)$. From the basic formulae (2.4) and (2.5), it can be easily shown that, in the frequency domain, this multiplication leads to a periodic reproduction of $X(f)$ (spectrum of $x(t)$) with 'period' $f_s = 1/h$. Now, if this 'period' is less than twice the upper frequency f_o of $x(t)$, the neighbouring replicas of $X(f)$ will overlap, so that the calculated FT of x_i will contain not only $X(f)$ but also the additions from $X(f - n/h)$, where $n = 1, 2, \ldots; -1, -2, \ldots$; in this way, frequencies higher than the sampling rate 'penetrate' into the sampled signal. This effect may sometimes be observed in old movies of the twenties and thirties, where, for example, the wheels of a running carriage of the 19th century are seen to be almost motionless: they apparently do not rotate at all, or may even rotate in the 'wrong' direction.

To avoid aliasing, without an unnecessary increase in the sampling rate, a low-pass prefiltering of the analogue signal x(t) is used generally before this signal is applied to the ADC. The cutoff frequency of this antialiasing low-pass filter is chosen in such a way as to preserve all the frequency components of x(t) that may be of interest for the given case. On the other hand, excluding all the 'nonrelevant' higher frequencies guarantees that they will not interfere with the 'relevant' lower-frequency components even if the sampling rate is only just above twice the value of the filter cutoff frequency. Therefore, this prefiltering may, in fact, provide a quite efficient preliminary data reduction and save computation costs.

Besides frequency-domain filtering, time-domain filtering, or temporal windowing, is used sometimes. Specifically, for a given signal, only values within a certain given time interval(s) are chosen for a subsequent analysis or interpretation. This may be done by multiplying the signal by a function of time, or window function, which is zero outside the above interval(s). This procedure is used, for example, to extract a 'direct' acoustic signal, for a source, from its mixture with 'echoes'; that is, those signals from the same source that arrive at the measurement point via various 'secondary' transmission paths.

An extremely important topic, related to the implementation of the Fourier transform, is the proper treatment of a real signal, which is available only within a finite time interval, whereas according to eqns. (2.4) and (2.5) the signal should be given everywhere within an infinite interval. This problem of truncation arises when the FT is calculated from the signal itself, or from its correlation function (with a random x(t)) to obtain the spectral density according to eqns. (1.11) and (1.12). The corresponding error in the calculated FT of x(t), which is sometimes called a 'leakage error', is clearly illustrated as follows. Let T be the available sample length of x(t), so that the lower and upper infinite integration limits in eqn. (2.5) are replaced by $-T/2$ and $T/2$, respectively. Formally, this is equivalent to the multiplication of x(t) by a 'natural data window' function w(t) such that $w(t) = 1$ for $|t| \leq T/2$ and $w(t) = 0$ elsewhere. It can then be easily shown that the FT of this truncated signal $x_T(t) = x(t)\, w(t)$ may be represented by the 'convolution integral':

$$X_T(f) = \int_{-\infty}^{\infty} X(\nu)\, W(f-\nu)\, d\nu \qquad (2.7)$$

where W(f) is the FT of the data window, w(t), and is the 'spectral window'; for the above 'natural' or rectangular data window, the corresponding spectral window is $W(f) = \sin(\pi f T)/\pi f$. Equation (2.7) clearly shows that, because of the truncation, other frequency components of X besides X(f) may contribute to the calculated FT, $X_T(f)$, unless W is a Dirac delta-function, so that $X(\nu) W(f-\nu) = 0$ for all $\nu \neq f$ and $X_T(f) = X(f)$. The latter ideal case is indeed approached if the sample length, T, is increased indefinitely. However, for any finite T, the FT of the truncated signal differs from that of the original one. It is this 'redistribution' in a frequency domain that is called leakage. To reduce the leakage error for a fixed finite value of T, various other windows w(t) (or corresponding W(f)) have been proposed by various authors [4,28,31]; they may indeed provide certain improvements in the calculated FT. In any case, decreasing the available sample length, T, of x(t) leads to the FT becoming more 'smeared'. This property is reflected in the 'Uncertainty Principle', which may be formulated as follows [29]: The maximal resolution in the frequency domain, which may be attained by any processing algorithm, is inversely proportional to the length of the processed sample of the signal. (Various shapes of time windows only influence the proportionality factor in this relation.)

The FT may be calculated by a digital computer very efficiently if certain symmetry properties of the integrand in eqn. (2.4) or eqn. (2.5) are used [28, 35]. This efficient algorithm is called a fast Fourier transform, or FFT. Thus, the first capital F in this abbreviation reflects the approach to calculations (which permits one to minimise the required number of elementary arithmetic operations) rather than the essence of the algorithm.

We outline now another important procedure for data processing, namely the extraction of the amplitude and phase of a narrowband process. Such a process, as shown in Chapter 1, may be represented in the form:

$$x(t) = x_c(t) \cos \Omega t + x_s(t) \sin \Omega t \tag{2.8}$$

or

$$x(t) = A(t) \sin [\Omega t + \varphi(t)] \tag{2.9}$$

where the amplitude A(t) and phase $\varphi(t)$ are both slowly varying compared with the mean frequency W of x(t), as well as the processes $x_s(t) = A \cos \varphi$ and $x_c(t) = A \sin \varphi$, which may be called the inphase and quadrature components of x(t), respectively (with respect to the signal $\cos \Omega t$). The simplest way of calculating all these slowly varying components for a given signal x(t) is to multiply x(t) by $\cos \Omega t$ and by $\sin \Omega t$ and then to low-pass-filter each of the products obtained with a cutoff frequency of the order Ω. The above products may be represented then as $x \cos \Omega t = x_c/2 + \ldots$, $x \sin \Omega t = x_s/2 + \ldots$, where the dots denote high-frequency components, proportional to $\cos 2\Omega t$ and $\sin 2\Omega t$. Thus, the above filtering operation directly yields the inphase and quadrature components of x(t), from which the amplitude and phase may subsequently be calculated as $A = (x_c^2 + x_s^2)^{1/2}$, $\varphi = \arctan (x_c/x_s)$. This method can be implemented both for digital and for analogue signals x(t), although in the latter case a pair of precise sine generators with adjustable phases of the output sine signal must be available. Moreover, whenever only the amplitude, A(t), is required, the algorithm may be somewhat less complex. Thus, the original signal x(t) may be directly squared and then low-pass-filtered, yielding the squared amplitude $V = A^2$, or envelope, since $x^2(t) = A^2(t) \sin^2 (\Omega t + \varphi) = V(t)/2 + \ldots$, where dots denote higher-frequency components of the filter output signal. For those x(t) that are not ideally narrowband, this technique may lead to some errors; and certain more sophisticated procedures, based on the 'Hilbert transformation', may be preferable [34].

Certain specific topics concerning random signals processing are now considered. These signals are assumed to be ergodic, so that time averaging according to Eqn. (1.7) may be used to estimate their statistical characteristics. Because of the finite length of the available record, or sample, of the signal, these estimates are always found to be different from the corresponding true characteristics. This random error may be described by its mean value (the so-called bias error) and the variance, which describes the random scatter of the finite-length estimate, obtained. If the former is zero, the estimate is called unbiased; it is always desirable to obtain unbiased estimates. On the other hand, any estimate should also be consistent; that is, the mean square of the zero-mean random part of its error should tend to zero with an indefinite increase in the sample length. For example, the estimate (1.7) of the mathematical expectation for a finite T has both these properties. For more sophisticated characteristics of a random process, the requirements of a small bias and a small variance of the random part of the error may sometimes prove to be contradictory, so that some tradeoff may become necessary. For example, when a histogram of a random process (or random variable) is generated to estimate its probability density (see Chapter 1, particularly Fig. 1.1), the bias error decreases with the decreasing size of the interval of the piecewise-constant approximation. On the other hand, this decrease may lead to a high random error, since the relative stay time of the signal within any narrow interval is small; thus, some tradeoff should indeed be sought. It may be added that two-dimensional or multidimensional probability densities of random processes (variables) are estimated essentially in a similar way; namely by introducing small intervals along each variable and calculating the relative stay times (relative frequencies of occurrence) within each cell of the whole domain of possible values of this random process or variable.

Similar tradeoff considerations are encountered in estimating the spectral density of a random process by the use of narrowband filters. If a signal is passed through a bandpass filter with a sufficiently small bandwidth, the variance of the filter output should provide an estimate of the value of the signal's spectral density at the central passband frequency (if the lower cutoff frequency is not zero). Thus, the spectral density can be measured either by simultaneously passing the signal through a set of such filters with various central passband frequencies, or by using a filter with an adjustable passband frequency and scanning in the frequency domain. This method is particularly suitable for analogue signals, and therefore a variety of commercial electronic analysers exist that are widely used for a preliminary 'rough' spectral analysis, to identify the most interesting frequency ranges, which should be preserved in the process of low- or bandpass filtering before the signal is converted into a digital form. An extremely important parameter of any such analyser is its bandwidth. As a matter of fact, the bias error of any such estimate of a spectral density is reduced by decreasing this bandwidth; however, this leads to a simultaneous increase in the variance of the corresponding random error. A quantitative analysis of these errors can be found in [4,31].

In digital processing, two other methods of spectral analysis may be preferable to that outlined above, since multiple digital filtering may lead to excessive computation costs. The first method is based on first directly estimating the autocorrelation function as $K_{XX}(\tau) = \langle x(t)\, x(t+\tau) \rangle$, on the basis of the general averaging rule (1.7). Then, the FFT of the estimate obtained is calculated with an appropriate 'window', which should lead to a consistent estimate. The other method is to calculate first the FFT of the signal itself (once again, with a smoothing window) and then to average the squared values of this FFT. In this latter method, an additional smoothing by averaging estimates over various subintervals of the whole sample is often used. The estimate of the autocorrelation function may be obtained from that of the spectral density by applying the inverse of the FFT procedure.

The above procedures for signal processing should be regarded as general-purpose ones. Many other procedures are studied in subsequent chapters in relation to specific VA diagnostic problems. We conclude this chapter with a general flow chart of a diagnostic system (Fig. 2.1), illustrating the basic sequence of operations. It should be regarded as a crude general illustration, with many variations being possible. For example, many systems with purely analogous signal processing do exist; they do not involve a digital computer (nor, therefore, an ADC), whereas the hardware for

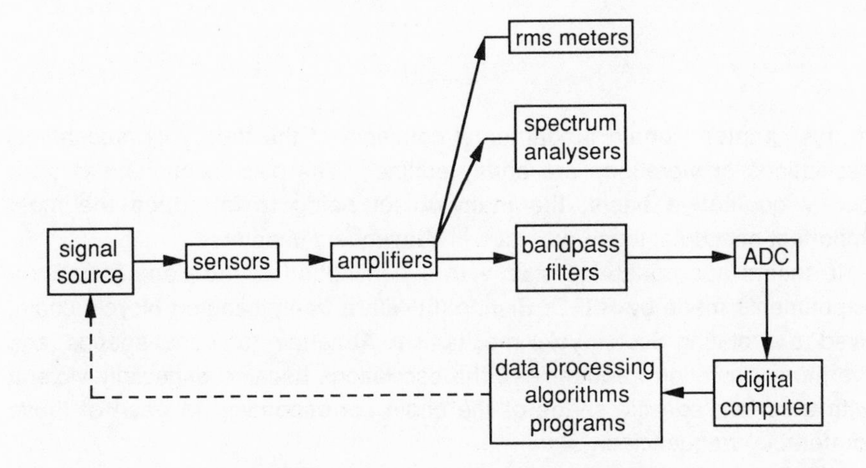

Fig. 2.1 General flow chart of a diagnostic system

processing may include other items besides spectrum analysers and rms meters. A dotted feedback line from the last block to the first one corresponds to the case of experiments with a controlled input signal (such as special vibration tests of structures); in this case, the results of signal processing may be used to make a proper adjustment of the parameters of the input signal, directly, in the course of the experiments. As explained in the Introduction, it is this last block, containing the algorithms and programs for signal processing, that is the main theme of this book.

Chapter 3 Basic Vibrational Phenomena and Their Analysis

In this chapter, certain fundamental concepts of the theory of mechanical oscillations or vibrations are briefly outlined. The discussion here is on a purely qualitative basis, the main object being to introduce the most important concepts; these are used in subsequent chapters.

It seems appropriate to start with a description of the demonstrational experiments made by R.E.D. Bishop [6] with a freely hanging bicycle chain, fixed to a rotating Scotch yoke mechanism. At certain rotational speeds (and therefore excitation frequencies), the oscillations became especially violent, with a certain specific shape of the chain corresponding to each of these 'preferable' frequencies.

When the rotation stopped suddenly, the chain still oscillated with this prescribed frequency and its corresponding shape (see Fig. 3.1, as reproduced from [6]), although these 'free oscillations' gradually decayed with time. Therefore, these frequencies and shapes may be regarded as the basic properties of the chain itself; as such, they are called the natural frequencies and natural shapes or modes, respectively. Oscillations in a certain mode, at the corresponding natural frequency, are sometimes called standing waves, as opposed to the so-called propagating or travelling waves, which are considered mainly in Chapter 4.

As a matter of fact, any deformable or flexible component possesses such natural frequencies and modes. The former are usually distinct ones. In sufficiently complex structures, however, certain distinct modes may occasionally have quite close natural frequencies. Moreover, sometimes 'multiple' natural frequencies, which are identical but correspond to different modes, may exist owing to certain symmetry properties of the system. Indeed, a freely hanging heavy cable of circular cross-section has no 'preferable' plane of oscillations, and therefore any mode of the type shown in Fig. 3.1 exists either in the plane of the Figure or in that normal to it; of course, with perfect axial symmetry of the cable, the corresponding natural frequencies of these modes are identical.

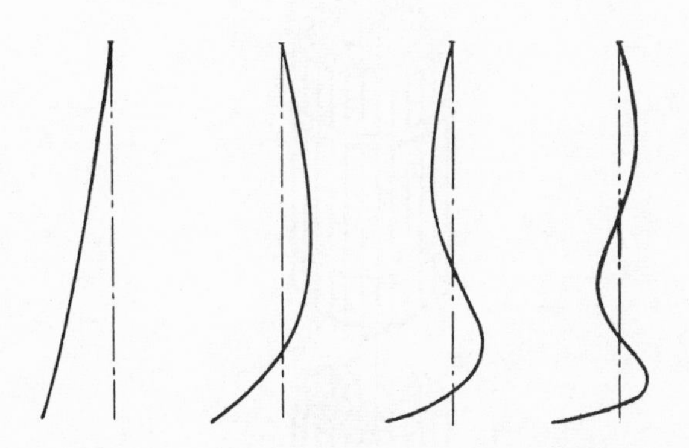

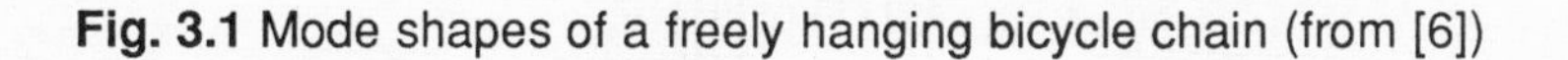

Fig. 3.1 Mode shapes of a freely hanging bicycle chain (from [6])

Of all the bicycle-chain modes shown in Fig. 3.1, the left-hand one is somewhat special, since throughout the oscillations in this mode the chain hardly changes from the straight-line shape that it possesses at rest, i.e. in a static equilibrium position (this linear shape would be achieved exactly if a sufficiently heavy weight were attached to the free lower end of the chain). This is a 'rigid-body mode', which does not involve any distortions of the system's shape. Other modes are called elastic, or flexible, ones since they correspond to significant deformations or changes of the system's shape.

An example of a system with both rigid-body and elastic modes is a ship which undergoes heaving and rolling (rigid-body!) motions in a rough sea, whereas its hull may simultaneously undergo elastic vibrations. Another example is a freely hanging fuel-bundle assembly of the channel-type nuclear reactor, as shown in Fig. 3.2. Several thin fuel rods are tied up in a bundle, by so-called spacers, with sufficient gaps for coolant flow, similarly to a quiver with arrows. Two such bundles are connected one to another by a common central rod (whose length, therefore, is a little more than twice that of a single rod in the bundle). The first rigid-body or pendulum-type mode of the assembly is shown schematically in Fig. 3.3(a). The second mode, with a higher natural frequency, as shown in Fig. 3.3(b), corresponds to rigid-body motions of each of the two bundles; such a mode may be identified as a rigid-body one because the part of the central rod between the two bundles is much more flexible than the bundle itself. It can also be seen that, whilst in

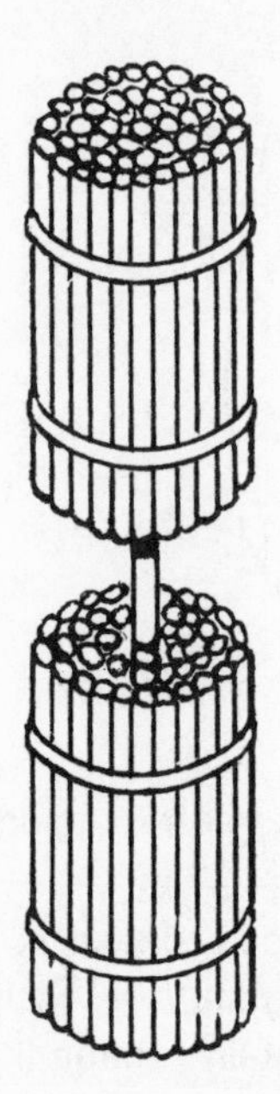

Fig. 3.2 Sketch of a fuel-bundle assembly of a channel-type nuclear reactor (not to scale)

Fig. 3.3(a) both bundles are inclined in the same direction (inphase modal displacements), in Fig. 3.3(b) they are inclined in opposite directions (out-of-phase modal displacements). It is noted that the assembly also possesses other modes, in which bending of the bundles is involved. These modes have much higher natural frequencies. However, they may be excited by impacts of the assembly against the walls of the channel tube within which the assembly is installed.

The last example clearly illustrates the fact that a system of interconnected rigid bodies (a 'lumped parameter system') may possess only a finite number of modes (in this case, two), and therefore also natural frequencies, depending on the number of bodies. On the other hand, a flexible or elastic system may have (theoretically) an infinite number of modes, although excitation of the very-high-frequency modes may sometimes be rather difficult. Moreover, as the size of a structure increases, its neighbouring

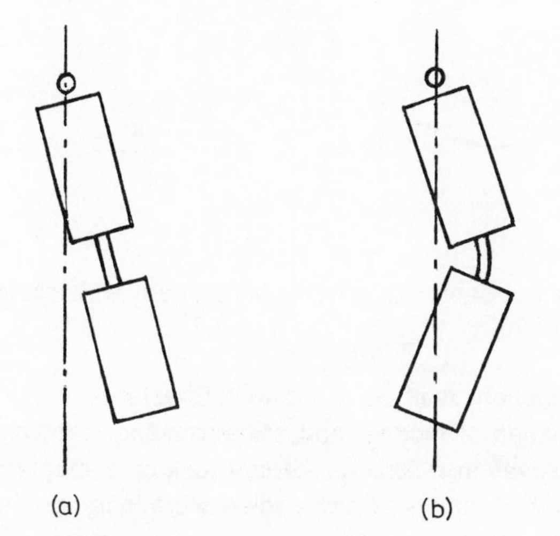

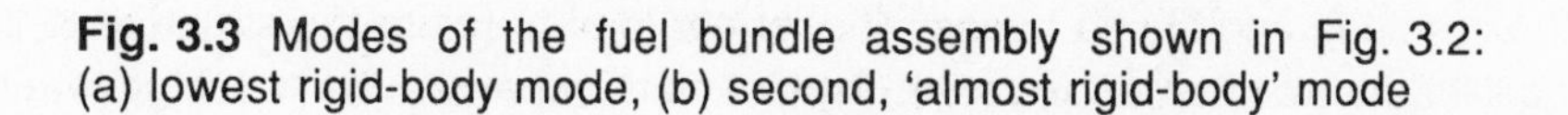
Fig. 3.3 Modes of the fuel bundle assembly shown in Fig. 3.2: (a) lowest rigid-body mode, (b) second, 'almost rigid-body' mode

natural frequencies become very close, making it appropriate to use the 'travelling-wave approach' to analysis, which is briefly outlined later. Both the elastic modes and their natural frequencies may be strongly dependent on the type of edge supports of the structure. Thus, Fig. 3.4 illustrates the fundamental (having minimal natural frequency) bending modes of an elastic beam for two cases. In case (a), both ends of the beam are clamped, or restrained completely, in such a way that both the transverse displacements and the rotations of these ends are zero. In case (b), these ends can rotate freely, whereas their transverse displacements are still zero ('hinged' , or pinned, ends). As a matter of fact, the natural frequency in case (a) is about 2.25 times as high as that in case (b), the reason being the additional restraint due to the clamping of the ends. For higher modes, such differences become smaller, and for modes with a high number of standing waves within the beam (or structure, in general) they may gradually diminish. Thus, if a decrease in a natural frequency(ies) is observed during the service life of a structural element or machine component, the possible implication is that something is wrong with its support(s); another possible reason is a crack

Fig. 3.4 Fundamental modes of a beam with clamped (a) and pinned (b) ends

within the component itself, as is shown in Chapter 9.

Other examples of modes, and corresponding standing waves, are the free surface waves in a liquid-propellant tank of a spacecraft (they may be excited by rocket motion and/or engine vibrations) and sounds (acoustic oscillations) in a room.

The most important forces that govern the free vibrations of a structure are the restoring and inertia forces. The former tend to return the structure to its equilibrium position after some displacement, whereas the latter generally prevent the structure from 'sticking' at this position when it is approached with a nonzero velocity. In the example of a freely hanging chain, the restoring forces are provided solely by gravity, whereas in most structures these forces are due to the stiffness of the structure itself (in cases where the elastic modes are considered) and/or its supports (for the rigid-body modes of the structure); such a restoring force for any mode is proportional to the characteristic displacement in this mode. The inertia forces of a structure are governed by its mass properties and are proportional to the accelerations of the structure. Both the elastic restoring and inertia forces may be evaluated relatively easily in terms of displacements and accelerations, respectively, according to the basic laws of mechanics. The solution of the resulting equations for free vibrations often provides sufficiently accurate predictions of both the natural frequencies and the modes of the system.

On the other hand, the restoring and inertia forces cannot by themselves provide a completely adequate description of free vibrations, since in the absence of any excitation source such vibrations always decay with time, whereas these two basic forces correspond to a system without energy dissipation. In Fig. 3.5, an example of such decaying oscillations in a single-

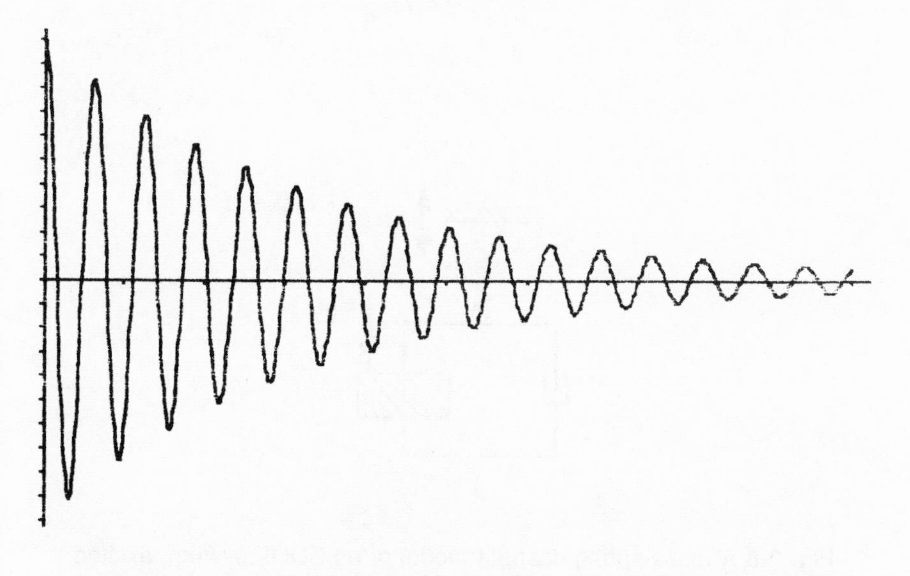

Fig. 3.5 Computer-drawn decaying free oscillations of an SDOF system

mode or single-degree-of-freedom (SDOF) system is shown, as obtained by computer simulation. Thus, some resistance forces are present in every system, although in many cases they may be quite small compared with the restoring and inertia forces. In such cases, this decay is found to be rather slow (as in Fig. 3.5). The source of this resistance, or damping, may be either internal friction within a structure due to inelastic deformations of materials, dry friction in joints etc. or its interaction with the environment, in which, say, friction and/or radiation of sound (acoustic waves) in air may be involved. In most cases, this damping cannot be predicted accurately from the basic laws of mechanics or physics and only very rough predictions are possible. Therefore, various methods for estimating damping from experimental vibration data are of the utmost importance; some of these methods are presented in subsequent chapters.

It is obvious that, since the free vibrations of any system always decay, sustained oscillations of the system may, in general, be possible only if some sort of external input excitation is applied. In the following, various possible types of such excitation are considered, at first for an SDOF system and then for a general multimodal case.

In Fig. 3.6, a basic mechanical model of an SDOF system is shown: namely a system, consisting of a mass, spring and dashpot, which provide inertia force, restoring force and damping, respectively. The system is excited

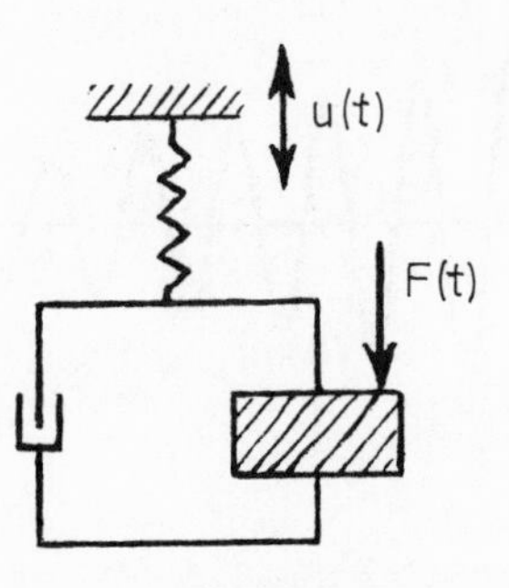

Fig. 3.6 A mass-spring-dashpot model of an SDOF system, excited by an external force F(t) and/or by a base acceleration ü(t)

either by an external force, F(t), or by a base motion, u(t). Numerous examples of the former case can be found in rotating machinery, where the source of such excitation may be shaft unbalance. The best example of the latter case of so-called kinematic excitation is the heaving motion of a car, travelling along a rough road (in this case, the system in Fig. 3.6 is viewed in a frame moving horizontally together with the car). In Fig. 3.6, both excitations are assumed to be independent of the system's state variables; that is, of the displacement and velocity of the system. This is not always the case. For example, the air loading on the wing of an aircraft may be dependent on the wing's displacements, giving rise to various 'aeroelastic' phenomena. These phenomena, however, are considered somewhat later, and we now focus on this first kind of excitation, due to time-variant external loads. The governing differential equation of motion of the SDOF system is*:

$$m\ddot{x} + 2c\dot{x} + kx = F(t) \tag{3.1}$$

Here, x(t) is a displacement of the system from its static equilibrium position,

* While it directly corresponds to the case of Fig. 3.6 with excitation force F(t), the same form of basic equation is obtained easily for the kinematic excitation case, with F(t) proportional to ü(t)

and the first, second and third terms on the left-hand-side represent inertia, damping and restoring forces, respectively (m is the mass and k the stiffness of the spring). Equation (3.1) may be rewritten in the form:

$$\ddot{x} + 2\,\alpha\,\dot{x} + \Omega^2 x = \frac{F(t)}{m} \quad ; \quad \alpha = \frac{c}{m} \quad , \quad \Omega = \sqrt{\frac{k}{m}} \tag{3.2}$$

where Ω is the natural frequency of the undamped system and α is a damping factor. In most mechanical systems without artificially introduced damping, the damping ratio α/Ω is small compared with unity (usually of the order $10^{-3} - 10^{-1}$).

The simplest law for the timewise variation of the excitation force is a harmonic or sinusoidal one, i.e. $F(t) = F_o \sin \nu t$. For this case, after a sufficiently long time from the start of the motion, when all transients due to the initial perturbations have decayed away completely, the resulting 'steady-state response' is also harmonic in time with the same frequency; i.e.

$$x(t) = A \sin(\nu t + \varphi) \tag{3.3}$$

Here, A and φ are the response amplitude and phase, respectively. Figure 3.7

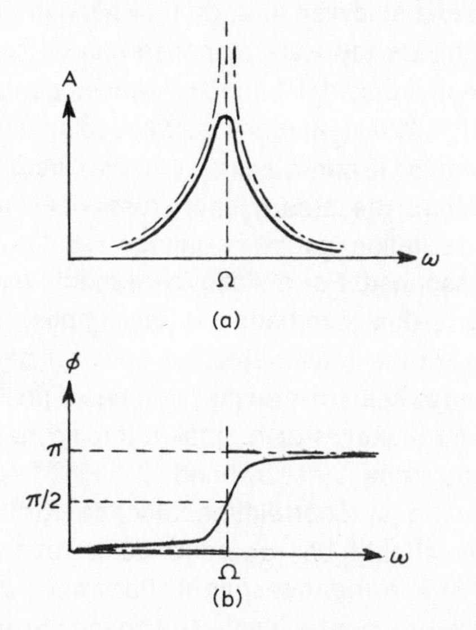

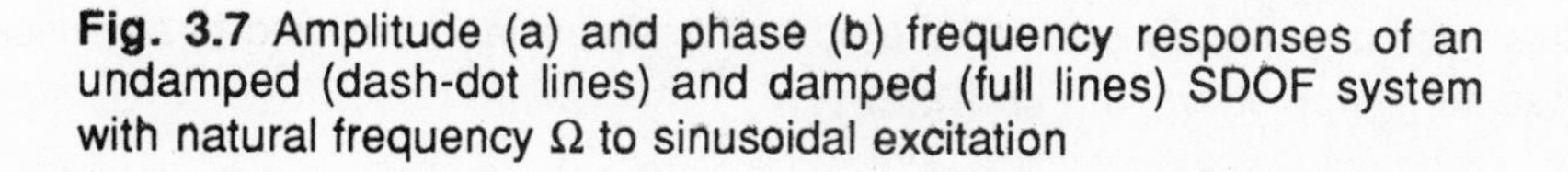
Fig. 3.7 Amplitude (a) and phase (b) frequency responses of an undamped (dash-dot lines) and damped (full lines) SDOF system with natural frequency Ω to sinusoidal excitation

illustrates their dependence on the excitation frequency, and particularly the so-called resonance in the case where the excitation frequency is tuned to the system's natural frequency ($\nu = \Omega$). The resonant condition may be defined also as corresponding to the case where the inertia and restoring forces completely cancel each other. The latter definition may be regarded as a general one, valid for SDOF models which account for various additional nonlinear and time-variant factors compared with the simplest model (3.2), and also for other types of excitation. It is clear that, at a resonance, the excitation force may be balanced only by damping, so that in the absence of the latter the response amplitude should grow in time indefinitely. With finite damping, the steady-state response amplitude is maximal, since any detuning from resonance implies that either the restoring force (if $\nu < \Omega$) or the inertia force (if $\nu > \Omega$) will also oppose the action of the excitation force. As for the phase, it is close to zero when α/Ω is small (and is zero exactly in the absence of damping) in the subresonant case ($\nu < \Omega$) and close to 180° in the postresonant case ($\nu > \Omega$). The former case may thus be regarded as being stiffness-dominated (small phase shift between the force and displacement, $\ddot{x}(t)$) and the latter one as inertia-dominated (small phase shift between force and acceleration, $x(t)$). A quite narrow frequency band in the vicinity of $\nu = \Omega$ with a bandwidth of the order 2α may be called the resonant domain; within this domain, the damping force is at least as important as either of the two other internal forces. In the undamped case, the phase-frequency response is a stepwise function (solid line in Fig. 3.7(b)).

This analysis of the steady-state response can be easily extended to the cases of general nonsinusoidal periodic, almost periodic, and/or random excitation force, $F(t)$. While a more detailed discussion of this topic is presented in Chapter 5, it can be simply stated here that, in view of the superposition principle, the steady-state responses to various harmonic components of the excitation can be calculated separately and then summed to give the overall response. For a stationary random excitation, which has a continuous spectrum, this summation in fact implies integration over the response spectrum. For a reasonably flat spectral density of the random excitation, the response spectral density $\Phi(\omega)$ has a peak at $\omega = \Omega$; i.e. at the system's resonant frequency. For a small damping ratio, this peak is very sharp, so that the response is narrowband: see Fig. 1.4(c), where both $\Phi(\omega)$ and the corresponding autocorrelation function $K(\tau)$ are shown. Such a response obviously should be regarded as a resonant one, since the contribution from these near-resonant harmonic components indeed dominates the overall response level. The response probability density in this case is generally close to the normal or Gaussian form, and in any case has the shape shown in Fig. 1.2(a).

Consider now another type of excitation which can lead to sustained periodic oscillations in certain systems in the absence of any time-variant external loading. Such oscillations are called self-excited or self-sustained. Of course, some external energy source is always required (otherwise the damping would lead to a complete decay of the oscillations due to any initial disturbances). However, in self-oscillatory systems, the environment provides only a constant energy source, whereas the system itself regulates its interaction with the environment, by some feedback mechanism, in such a way that it is acted on by a periodic force with an appropriate phase shift with respect to the system's velocity. Many self-excitation phenomena may be found in structures interacting with a fluid flow, since the loading from fluid flow may, in general, be dependent on the structure's motion. The best-known example is probably the flutter of an aircraft wing; this phenomenon is considered later. Many systems with dry friction may be prone to self-excitation; well known examples include the squeal of car brakes and the sounds of a violin string excited by a bow.

We do not discuss here the details of these interactions, which may lead to self-excitation (see many interesting examples in [6,9,23,41]), and consider only their possible net effect for an SDOF system (3.2) with $F(t) \equiv 0$. Suppose that somehow the apparent damping factor α becomes negative, so that the 'resistance' force acts in the same direction as the system's velocity, rather than in the opposite one. It should be expected that, while the restoring and inertia forces may still cancel each other, this 'negative resistance' force will lead to a growth of oscillations at the system's natural frequency, due to any initial disturbance. As an example, the fuel-bundle assembly shown in Fig. 3.2 may be mentioned. An upward coolant flow with a sufficient velocity will lead to self-excitation of the lowest (rigid-body) mode (shown in Fig. 3.3(a)) provided that an apparent negative hydrodynamic damping exceeds the positive structural damping.

In such a situation, nothing can prevent the gradual growth of the response amplitude until there is a complete failure of the structure. This simply means that the linear model (3.2) does not describe adequately the behaviour of SDOF systems, in which periodic self-excited oscillations (or 'limit cycles') are possible. The nonlinearity of the restoring force cannot limit the growth of the response amplitude due to a negative-resistance force; such a limit can be attained only if the damping becomes positive at sufficiently high response amplitudes. Specifically, let the equation of a self-oscillatory SDOF system be:

$$\ddot{x} + f(A^2)\,\dot{x} + \Omega^2 x = 0 \quad , \quad A^2 = x^2 + \frac{\dot{x}^2}{\Omega^2} \tag{3.4}$$

Then, periodic limit-cycle oscillations are possible provided that $f(A^2)$ is positive for all sufficiently high amplitudes, A, and negative within a certain finite range of A. If this range includes the origin $A = 0$, we have the above case of a negative-damping dynamic instability of a linear model ($2\alpha = f(0)$); such a self-excitation is a 'soft' one. On the other hand, let α be positive, but $f(A^2)$ be negative within a certain range $A_1 < A < A_2$, where A_1 is strictly positive. Then self-excitation is also possible, but it requires a high initial disturbance, sufficient to bring the system within this amplitude range of negative damping; this is 'hard' self-excitation. Similar behaviour may be observed where a small damping force depends on both x and $\dot{x}$ (not just on A^2); in this case, the conditions for self-excitation depend on the averaged-over-the-period ($2\pi/\Omega$) damping factor.

In the above example of a fuel bundle in an upward coolant flow, the displacements of the bundle are restricted by the tube wall that surrounds the bundle. This tube wall in fact provides a strongly nonlinear restoring force, with the additional stiffness becoming essentially infinite after each impact. For limit-cycle oscillations to be possible, however, it is more important that there is sufficient additional damping due to the inelasticity of the impacts (the velocity of rebound is usually less than that of impact); otherwise, a negative damping would lead to an unrestricted growth of the impact velocity.

Consider now yet another type of excitation, which involves time-variant external forces, but not in the same way as in the above case of external excitation. Specifically, these forces lead to timewise variations of the system's parameters — stiffness, mass or damping. Such an excitation is 'parametric', and in certain conditions it may lead to violent oscillations of a structural element or machine component. For systems with restoring forces due to gravity, parametric excitation may become possible owing to environmental oscillations that lead to timewise variations in the apparent gravity acceleration. Thus, vertical oscillations of the suspension point of a pendulum lead to an additional time-variant 'stiffness' component, which is proportional to the acceleration of the suspension point. For example, the purely vertical or heaving motion of a ship in a rough sea may influence its rolling motion via such variations in the effective gravity. In thin structural elements, parametric excitation of lateral vibrations may be induced by a time-variant axial force, since the bending stiffness of the element is generally dependent on this force.

The latter point is illustrated in further detail in Fig. 3.8. Here, a beam with pinned ends is shown, which is loaded by a time-variant axial force P(t) and oscillates at its first bending mode with a corresponding natural frequency Ω. The dashed lines show the (symmetrical) shapes of the deflected beam in two extremal positions, and the axial force clearly provides an additional

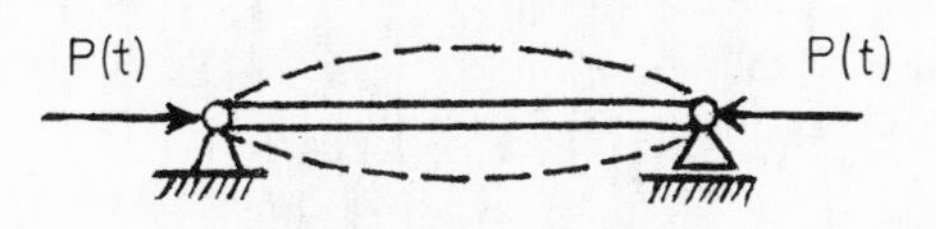

Fig. 3.8 Lateral vibrations of a beam with a time-varying axial compressive force. Dashed lines illustrate possible extremal positions of the beam's neutral axis within a given oscillation cycle

bending moment in the beam. It is obvious that the most favourable situation for excitation is where, at the time instants of these extremal positions in Fig. 3.8, the axial force is compressive and its magnitude is maximal. Therefore, the time-variant axial force should be periodic with a period equal to the halfperiod of the oscillations (the time interval between two successive instants for reaching the dashed positions in Fig. 3.8).

Another example is a control rod of a nuclear reactor — a slender vertical body with a laterally fixed lower end, which carries a graphite block. Graphite is a good neutron moderator and thus is used widely in feedback control systems of nuclear reactors, where it is required to provide the necessary variations in the absorption level. Therefore, the control system assigns longitudinal motions to the control rod. In certain structures, these motions may be implemented by a step-drive motor. The latter provides an intermittent axial force, and the corresponding longitudinal acceleration of the control rod is shown in the lower trace of Fig. 3.9 [36].

This axial force, however, can induce violent lateral vibrations of the slender control rod with a frequency one half of that of the excitation, as can be seen from the upper trace in Fig. 3.9. This is a phenomenon of parametric excitation, although the response trace also contains a component with the same frequency as that of the excitation. The probable reason for this direct external excitation is a certain inevitable misalignment of the rod, so that the force of the step-drive motor is not perfectly axial and also contains a certain lateral component.

The governing differential equation of a linear SDOF system with parametric

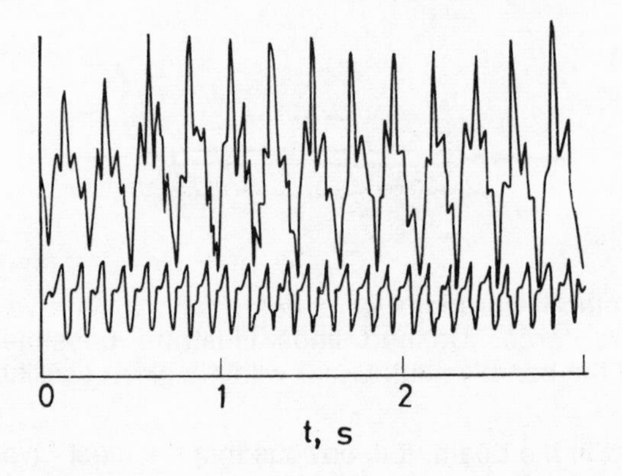

Fig. 3.9 Lateral vibrations of a control rod of a nuclear reactor (upper trace), excited by periodic variations in longitudinal acceleration due to a step-drive motor (lower trace). Response components with the same and with the doubled period of excitation are induced by the lateral and axial components of the excitation force, respectively

excitation due to stiffness variations may be written as (cf. with eqn. (3.2) with $F(t) \equiv 0$):

$$\ddot{x} + 2\alpha\dot{x} + \Omega^2 x\,[1 + \lambda\,\xi(t)] = 0 \qquad (3.5)$$

Assume first that $\xi(t) = \sin 2\nu t$, where ν is close to Ω. Mathematical analysis of this equation shows that the above qualitative considerations are correct: if λ is sufficiently high, then the obvious or trivial solution $x = 0$ (corresponding to the straight undeflected shape of a beam or control rod, in the above examples) becomes unstable and, after any small initial disturbance x_o, the response amplitude will grow indefinitely. The most favourable condition for instability is indeed that one which corresponds to the exact tuning to this 'main parametric resonance': the critical value of λ at the stability boundary is found to be minimal when $\nu = \Omega$. The full stability map of the boundaries for system (3.5) with $\xi(t) = \lambda \sin 2\nu t$ in the plane Ω/ν, λ — a so-called Ince-Strutt chart — is presented in Fig. 3.10. Here, a whole set of instability zones can be

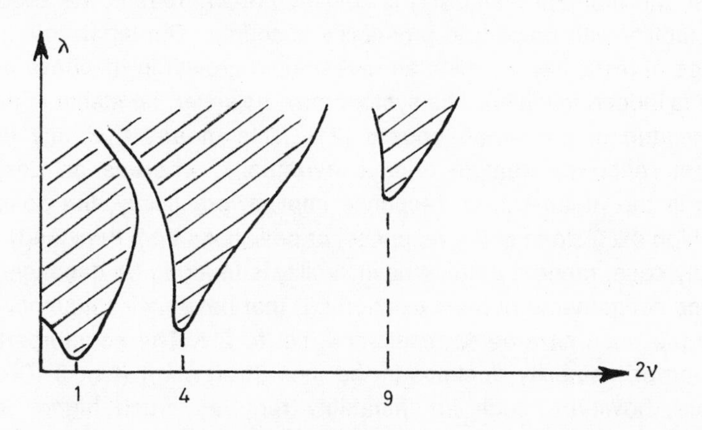

Fig. 3.10 Ince-Strutt chart for the system (3.5) with $\xi(t) = \lambda \sin 2\nu t$ (instability zones are hatched)

seen (hatched), where higher zones correspond to excitation frequencies such that the values of Ω/ν are of the order 4, 9, 16 etc. The width of the main zone ($\nu \approx \Omega$) is of the order 2α.

As in the case of negative damping, a linear model is also found to be inadequate to describe the steady-state response to parametric excitation. Indeed, whenever the net effect of the excitation prevails over that of damping, so that the linear system becomes unstable, no other force in the system seems able to restrict the growth in amplitude. However, with parametric instability, such a restriction may be provided not only by nonlinearity of damping, but also by nonlinearity of the restoring force. The reason is that the latter type of nonlinearity implies a variation in the mean stiffness, and therefore in the system's natural frequency, with an increase in the response amplitude; thus, it controls the source of instability by detuning the system from parametric resonance at sufficiently high amplitudes.

Parametric instability of an SDOF system is possible also in the case where the excitation $x(t)$ in eqn. (3.5) is a random rather than a periodic process. Stability analysis for this case is rather sophisticated, first of all because of the various possible definitions of stability for such cases, or of stochastic stability. Indeed, whereas the 'usual' deterministic concept of instability implies simply an unrestricted growth in response, after any initial perturbation of the system from its steady state, in the random case various indices may be introduced to describe how close the system is to the steady state. For example, the mean-square response may be used as such an index (or some other response moment, in general) or the probability of exceeding a certain response level. In the first case, the 'mean-square instability' (or, in general, the moment instability) is considered, whereas in the second case the instability with respect to probability is sought. The latter, in principle, is that type of instability in which an unrestricted growth in response amplitude in time is indeed involved. The system may, however, be stable in this sense but unstable in the mean square [21]. This means that any individual transient response sample decays eventually, whereas an unrestricted growth in the mean-square response implies, practically, the possibility of rather high excursions of the response, or deviations from the steady state.

In any case, random parametric instability is found to be dependent mostly on those components of the excitation $\xi(t)$ that have their frequency close to that of the main parametric resonance, i.e. to 2Ω. The counterparts of the higher-order instability domains in the Ince-Strutt chart (Fig. 3.10) are also possible; however, such an instability requires much higher levels of parametric excitation (high $\Phi_{\xi\xi}(4\Omega)$ etc.).

A steady-state response is possible in the case of a broadband random parametric excitation, provided, however, that the stiffness nonlinearity is of the 'hardening' type; i.e. the effective stiffness increases with the response amplitude. The latter in this case may have probability densities of the types shown in Fig. 3.11, with curves I and II corresponding to the cases of relatively low and high excess of the 'threshold' parametric excitation level, respectively.

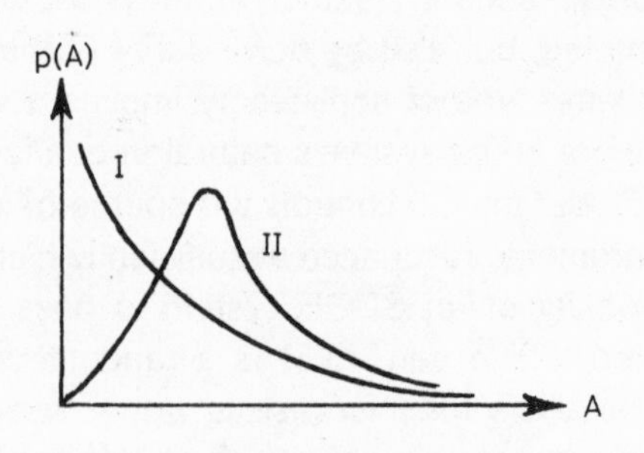

Fig. 3.11 Stationary probability densities of the response amplitude of an SDOF system with random parametric excitation and nonlinear stiffness and/or damping for a small (curve I) and high (curve II) excess of stability threshold

It seems appropriate at this stage to consider certain effects of a nonlinear restoring force, for the case of a purely external excitation. Thus, consider the following generalisation of eqn. (3.2):

$$\ddot{x} + 2\alpha\dot{x} + f(x) = \frac{F(t)}{m} \quad , \quad f(x) = \Omega^2 x + \gamma g(x) \tag{3.6}$$

In Fig. 3.12, 'backbone' curves show how the frequency ω of free undamped

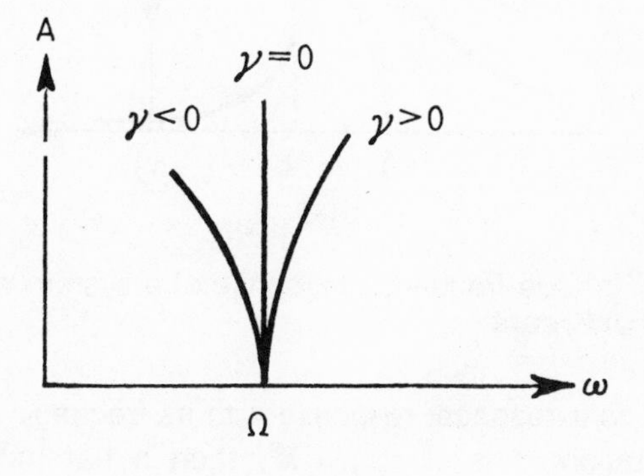

Fig. 3.12 'Backbone' curves of SDOF systems with 'hardening' ($\gamma > 0$) and 'softening' ($\gamma < 0$) nonlinearities

oscillations (for the case $F(t) \equiv 0$, $\alpha = 0$) depends on their amplitude A, in the case of a smooth nonlinear part, $\gamma g(x)$, of $f(x)$. Naturally, ω increases with A if $\gamma > 0$; i.e. if the apparent stiffness is increasing. Such is the case, for example, when a straight beam or plate vibrates laterally with moderate deflections, so that the axial tension provides an additional bending stiffness. The case $\gamma < 0$ is that of a 'softening nonlinearity', and ω is decreasing with increasing amplitude; the most obvious example of such a case is a simple pendulum: with increasing inclination angle, the restoring torque due to the gravity force increases less rapidly.

The 'backbone' curve of an SDOF system with $\gamma > 0$ is also presented in Fig. 3.13 (by a chain-dot line), together with the response amplitude curve of system (3.6) with $F(t) = a \sin \nu t$. Of course, contrary to the linear case, the steady-state solution of eqn. (3.6) is generally not a simple harmonic of type (3.3), and may contain higher harmonics of the excitation frequency ν, or 'superharmonics'. Moreover, in certain cases, 'subharmonic response' is

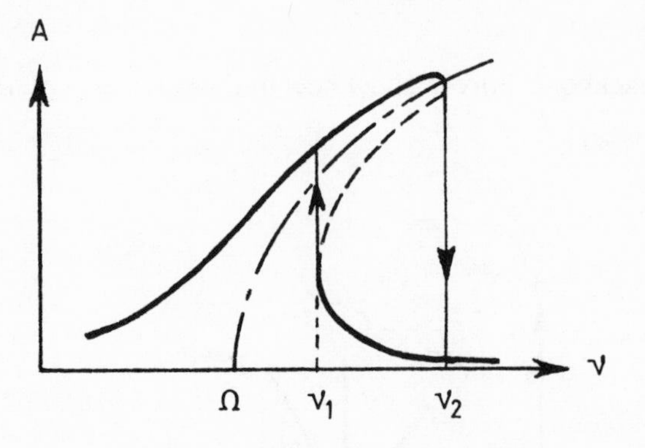

Fig. 3.13 Amplitude-frequency response of a system with a smooth hardening nonlinearity

possible, i.e. a near-resonant response with its frequency an integer divisor of that of excitation; thus, if $g(x) = x^3$, then a harmonic excitation with frequency $\nu \approx 3\Omega$ induces such subharmonic resonance with a frequency $\frac{1}{3}\nu$ of the corresponding response component. Such subharmonics are used in a method of crack detection, outlined in Chapter 9. In Fig. 3.13, A is the amplitude of the fundamental harmonic of the response with excitation frequency ν. The curve is seen to be multivalued within a certain range of excitation frequencies $\nu > \Omega$. The middle branch of $A(\nu)$ in this range $\nu_1 < \nu < \nu_2$, as shown by the dashes, is unstable, and thus such a steady-state response cannot be achieved in any physical experiment. The upper and lower branches are stable, and they are achieved in tests with decreasing and increasing excitation frequency, respectively. At the points ν_1 and ν_2, where the curve $A(\nu)$ has vertical tangents, the 'jumps' in the response amplitude are usually observed, as shown by the arrows. In the absence of any additional disturbances, the response with the initially assigned amplitude (at the upper or the lower branch in case $\nu_1 < \nu < \nu_2$) will be preserved indefinitely. However, an additional random excitation can lead to spontaneous transitions between these stable branches. Similar transitions are also observed when $F(t)$ is a narrowband random process with its centre or mean frequency within the above range of multivalued A.

The above properties of nonlinear systems may be important for vibration control methods based on detuning the excitation frequency from resonance. Indeed, while in the linear case it is sufficient to ensure that the detuning $|\nu - \Omega|$ is much higher than the 'natural' system's bandwidth, 2α, for nonlinear systems the resonance domain may prove to be much wider, leading to more stringent and complex requirements for the necessary detuning. Such was the case with the control rod of a nuclear reactor after it had been installed within its 'channelling tube' (the traces in Fig. 3.9 were obtained during 'free' tests, without this tube). The tube wall provided essentially an infinite stiffness, and thus led to such a widening of the various parametric resonance domains that they almost overlapped, leading to rather narrow ranges of allowable frequency for the step drive. It may be added that this broadening of a resonant domain, owing to the nonlinearity of the restoring force, may also have important implications for diagnostics, as shown in subsequent chapters. Moreover, the above specific nonlinear effects may be used sometimes for crack detection in machine components, using vibrational response data (see Chapter 9).

Consider now multi-degree-of-freedom (MDOF) or multimodal oscillations, starting with the case of a linear system with purely external excitation. Using the superposition principle, such oscillations can be represented in general as a sum of various modal responses. If the system is free from 'nonconservative loads' (this concept is introduced later on), all these modal responses are uncoupled*, and each of them is governed by a differential equation of the form (3.2). For a 'continuous system', i.e. for a system with spatially distributed mass and stiffness properties, the modal or generalised forces depend on the spatial distribution of the loading. For example, the most effective excitation of a mode is provided by a concentrated force applied at the 'antinode' of this mode, i.e. at the point (or line in a two-dimensional structure) with maximum displacement (e.g. at the midspan for modes shown in Fig. 3.4). On the other hand, a concentrated force at the nodal point of a certain mode (i.e. a point with zero displacement) produces no excitation for this mode.

Thus, estimation of the steady-state response of a system to any given periodic or random excitation is straightforward, as long as the system's parameters are known. The response spectral density for a random loading usually contains a set of peaks, corresponding to the various natural frequencies of the system. For a pair of close natural frequencies Ω_1, Ω_2, the response spectral density in the corresponding resonant domain may have the shape shown in Fig. 1.4(d); the corresponding narrowband resonant

* Provided that damping is of a certain specific type; this assumption can usually be adopted whenever no major crosscoupling effects due to damping are observed

component may have an autocorrelation function, exhibiting the well known 'beat' phenomenon. Later in this chapter, another extreme case is discussed where all the neighbouring natural frequencies are so close that the response spectrum becomes quite smooth.

The undamped modes of a system are uncoupled only as long as the system is loaded by forces with fixed orientations. If, however, a force changes its direction depending on the system's response — such a force is called nonconservative — it may lead to interactions between the various modes and eventually to an instability of the system [9]. In many cases, such a nonconservative loading of the structure is provided by the forces from a fluid flow. The simplest example is probably that of the 'follower' force. In Fig. 3.14, a cantilever beam or column is shown loaded by a compressive force P

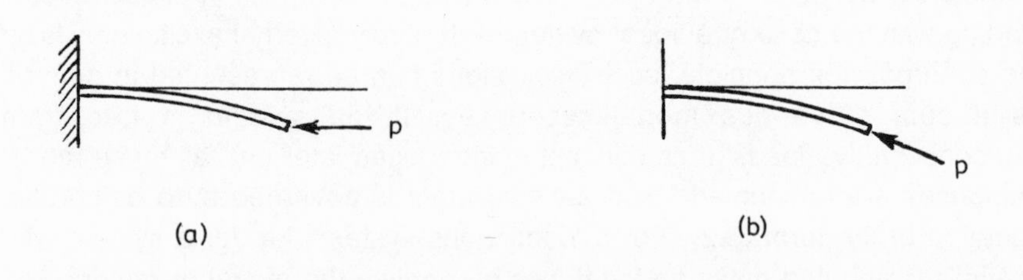

Fig. 3.14 A cantilever beam with 'dead' (a) and 'follower' (b) compressive loading at its free end

at its free end. In case (a), this force is always directed along the undeformed neutral axis of the beam. It is well known that such a 'dead weight' force of sufficient magnitude may lead to static instability or buckling of the beam. On the other hand, in case (b), the force p is always directed along a tangent to the axis of a deformed beam; such a force of sufficient magnitude may lead to dynamic instability of the beam. Thus, a sufficiently high flow of water through a hose, lying on the ground, may induce its violent oscillations: the hose wriggles like a snake owing to this follower force from the liquid jet.

Dynamic instability due to nonconservative loading is inherently a two-degree-of-freedom (TDOF) phenomenon, although in principle other modes may also participate in the resulting motion. Increasing the level of a nonconservative loading (in the above example of a flow velocity through a hose) may lead to the mutual approaching of a pair of natural frequencies of interacting modes, so that eventually they may coalesce. This point corresponds to the stability threshold of the undamped system, and after a slight additional increase of nonconservative loading, required to overcome damping, oscillations may start, with growing amplitudes of both modal responses.

Perhaps the most well known phenomenon of such a type is flutter. In the late thirties and forties, 'wing flutter' had become a source of many catastrophes with aircraft. In this phenomenon, which may be observed after a certain critical flight speed is exceeded, coupled bending and torsional vibrations of the wing are involved with increasing amplitudes; sometimes they may lead to fracture of a wing after a dozen seconds. An interesting eye-witness description of this phenomenon and its qualitative analysis are presented in [41].

Sometimes flutter may not be so dangerous, since nonlinearities of stiffnesses and/or damping may restrict the growth of the response amplitudes, leading to periodic limit-cycle oscillations with moderate amplitudes. Such is the case, for example, with the flutter of compressor blades in aircraft gas-turbine engines and the 'panel flutter' of aircraft structures (where the skin may oscillate similarly to a flapping flag in the wind). Nevertheless, such a 'mild' flutter may also be of concern to engineers, since it may lead to a fatigue failure of the structure.

Another well known example of dynamic instability due to nonconservative loading is the lateral instability of a car or railway vehicle travelling with a sufficiently high speed. In this case, the two basic degrees of freedom, related to the lateral translation of the vehicle mass centre and to its angular motion (yawing) around it, are coupled through a sophisticated mechanism of interaction between the wheels and the track.

A certain specific TDOF phenomenon is possible also in multimodal systems with parametric excitation. Thus, if certain modes are crosscoupled through a common parametric excitation, then in certain conditions they may become dynamically unstable. Thus, when the excitation is harmonic in time, this will be the case if its frequency equals the sum or difference of the natural frequencies of the interacting modes. This is a 'combinational resonance', and it may be observed for a control rod of a nuclear reactor as described in relation to Fig. 3.9. Another example is that of a helicopter rotor blade, in which 'coupled flapwise-torsional oscillations' may be excited [33]; in this case, the parametric excitation may be both periodic (due to the translational motion of a rotating blade in forward flight) and random in time (due to horizontal turbulent gusts).

We conclude this chapter with a brief outline of a quite different treatment of multimodal oscillations. Indeed, the above approach, based on a superposition of distinct modal responses or 'standing waves', may become impracticable for structures with high modal densities, i.e. for cases of high values of $\alpha\, n(\Omega)$. Here, α is a representative modal damping, whereas $n(\Omega)$ is the mean number of natural frequencies per unit frequency range, so that $1/n(\Omega)$ is the mean separation between a pair of consecutive natural frequencies. Therefore, the basic nature of the systems' response in cases where $\alpha\, n(\Omega) >> 1$ may be quite different from that in the case of superimposed distinct standing waves: as mentioned already, all neighbouring resonant domains may become so strongly overlapped that the amplitude frequency response or, say, the spectral density of the system's response to external random excitation is found to be quite smooth, without any distinct resonant peaks. Such cases may be treated by specific methods, based on the high numbers of modes involved [9,34,40]. The baseline extreme case for all these approaches is just the opposite to that for the modal decomposition, or standing-wave, approach. Thus, a 'continuous spectrum' of natural frequencies is considered (rather than that of the distinct natural frequencies of the various modes), typical for systems with distributed parameters, which occupy infinite domains. For systems of such a type, a travelling-wave approach seems to be more appropriate, and is considered in the next chapter.

Chapter 4
Travelling-Wave Responses and Certain Problems of Their Interpretation

Consider a long horizontal rope or string with both its ends fixed in such a way that the string is stretched. If the string is struck laterally at a certain point, a local zone of flexure, or 'bulge', is produced, which is then split into a pair of equal bulges, propagating to the right and to the left of the initial point of perturbation. These waves may be called forward and backward ones (if a positive direction along the string is chosen). Each wave travels with a constant velocity. Moreover, each signal or bulge preserves its shape and only its amplitude may decay gradually along the string. This important property implies that all the various frequency components of the signal propagate with the same velocity. Such media, with constant (independent of frequency) propagation velocities, are called nondispersive. On the other hand, dependence of the propagation velocity on the frequency of a harmonic-in-time wave is called dispersion. It is obvious that this effect should in general lead to a distortion of a signal's shape along its path.

The best-known examples of nondispersive waves are sound or acoustic waves (small-amplitude pressure and velocity oscillations in fluids) with the values of the constant propagation velocity or speed of sound (they depend only on the properties of the acoustic medium) being about 330 m/s for air and 1500 m/s for water, and axial or longitudinal stress waves in an elastic rod with the value of propagation velocity being about 5000 m/s for steel. The best-known example of dispersive waves is that of flexural or bending waves in elastic beams or thin plates, with their propagation velocity being proportional to the square root of the frequency. (This law is true, however, only as long as the wave-length is very high compared with the lateral dimensions, such as the thickness of the plate; waves with higher frequencies and correspondingly smaller wavelengths are governed by more complex laws, which lead to finite propagation velocities even at arbitrarily high frequencies.)

As mentioned already, the wave or 'signal' in a stretched wire is attenuated along its path owing to inevitable propagation losses. Therefore, it may decay almost completely before it reaches the nearest fixed end of the rope, or at least the reflected wave will not reach the region of interest. In such cases, the system or medium with the propagating waves may essentially be regarded as infinite; for example, the waves excited on a surface of a pond by a dropped stone may very well be regarded as purely propagating ones, without taking into account any reflections from the beach. Such an approach, which may greatly simplify both the analysis and interpretation of the measured responses, may prove to be quite natural in numerous applications, such as seismology or ocean acoustics. Of course, for structures or machines the possibilities for such a treatment are somewhat limited; usually, in view of the finite structure's dimensions, it may only be appropriate, if at all, at rather high frequencies.

Whenever a wave in the string or rope reaches one of the fixed ends, it reflects from it and then starts to travel in the opposite direction. Of course, such reflected waves, or 'echoes', may be accounted for, both in design analyses and in procedures for response data processing and interpretation. However, this approach may become impracticable in cases of multiple reflections, where too many reflected waves are present; in fact, a standing-wave pattern is the result of the interference and superposition of these multiple travelling waves. In any case, nevertheless, both the standing-wave and travelling-wave approaches may be important for the vibroacoustical diagnostics of machines and structures, as will be seen from various examples presented later in this book. Therefore, in the Chapter, certain basic wave phenomena and their analysis are briefly outlined, together with certain simple algorithms for the identification of the properties of a medium or structure from measured travelling-wave response data. One-dimensional waves are mainly considered, since any thorough treatment of two- and/or three-dimensional effects would be too complex for this book. However, the concept of wave diffraction is briefly considered.

The waves produced by certain initial disturbances, such as those in the above example of a stretched string, may be called free waves, similarly to free oscillations. In complex structural systems, various types of wave usually do exist. In general, they may be coupled, so that an exchange of energy between propagating waves of different types may be possible. Consider, for example, a thin-walled pipeline filled with fluid; this is indeed a structural system for which the travelling-wave approach may often be appropriate. The possible basic types of deformation, which may propagate as waves, are as follows:

(i) Axial or longitudinal, with propagation velocity c_l = 5000 m/s for steel.
(ii) Torsional, with propagation velocity c_t about $c_t = c_l/1.6$.
(iii) Bending or flexural, with propagation velocity $c_f = c_f(\omega)$; in general,

bending may be possible in two mutually perpendicular planes.
(iv) Pressure wave in a fluid within the pipe, with a propagation velocity generally less than the speed of sound c_0: because of the additional elastic compliance of the tube walls, this wave is accompanied by an axisymmetric or 'breathing' elastic wave in the pipe wall. In the extreme case of highly flexible pipe walls, such as those made of rubber or tissue, the propagation velocity of such a 'pulse' wave is governed by the wall's elastic properties and may not be dependent on the speed of sound in the fluid.

It may be added that, in pipes of large diameter, so-called 'shell' modes may also be possible, with multiple circumferential waves. In such a case, the structure, or 'waveguide' as it is called sometimes in acoustics, may in general be two- or three-dimensional. However, these waves still only propagate along one axis, whereas waves in the lateral direction may be regarded as standing ones. Thus, for each separate mode, corresponding to a certain standing wave in the lateral direction(s), one-dimensional propagation is still considered. For simplicity, however, these 'shell' modes are ignored here, and the beam-type model of the pipeline is assumed to be appropriate. Furthermore, the velocity of the flowing fluid in the pipe may be assumed to be small compared with c_0, otherwise forward and backward pressure waves would propagate with different velocities. However, this effect of a 'splitting' of the propagation velocity in moving bodies or media may be significant, say, for the bending vibrations of a rotating bladed disc of a gas turbine or compressor: when such a wave propagates along a circumference in the direction of rotation, the rotational speed is added to that of a wave in the motionless disc, whereas for rotation in the opposite direction the rotational speed is subtracted from this 'natural' propagation velocity.

The above four types of wave in a pipe are uncoupled only as long as the pipe is straight and the wave amplitudes are small. In a curved pipe, all these modes of deformation are coupled, so that an energy exchange between the propagating waves is inevitable. In coupled modes, one cannot, in general, clearly identify waves of different types in their pure form; at best, only the predominant modes may sometimes be ascribed to some waves when the coupling is sufficiently strong and thus has a significant influence on the propagation velocities. Furthermore, when a wave of a certain type is registered in such a 'composite waveguide' with various types of free wave, one can expect that waves of other types, with lower propagation velocities, will soon arrive. For example, an earthquake in the Pacific Ocean area may often induce not only an elastic wave in the earth, but also a strong sea wave. The latter, known as a tsunami wave, with heights up to 40–50 m near the coast, may be extremely devastating for coastal areas. It may propagate with velocities up to 800 kilometres per hour, which is quite fast for sea waves, but, of course, quite slow compared with waves in the earth (sea bottom). Therefore, a registered seismic signal may be regarded sometimes as a warning of the imminent arrival of a strong tidal or tsunami wave. However, in order to avoid numerous false alarms, one should be able to discriminate, more or less reliably, between ground motions that are and that are not accompanied by the tsunami wave; this problem is far from simple.

The last example, in addition to that of the fluid-filled pipeline, also clearly illustrates the concept of a composite waveguide. Each uncoupled part or subsystem of a composite waveguide has its own propagation-velocity characteristic: in the latter of the above examples, it is the speed of sound in the fluid and the 'dispersional relation' for a given mode of the shell *in vacuo* (the dependence of its propagation velocity on frequency). Whenever these subsystems are coupled within a composite waveguide, the propagation characteristics naturally become shifted from their uncoupled values. This shift is dependent on the coupling conditions at the interface(s) between the subsystems.

Moreover, these interface conditions may govern the proportions in which the total energy or power flux of the wave is shared between the various subsystems. A knowledge of such a sharing may be very important for vibration control. For example, the use of dampers or damping layers etc. for the vibration control of a fluid-filled pipeline will be effective only if a significant part of the energy propagates along its walls, rather than in the fluid. In [8], simple formulae were derived relating this energy or power flux sharing, for simple harmonic waves in a composite waveguide, to the interface condition(s). Calculations of energy and power flux sharing were made in [8] for the example of a fluid-filled shell, as functions of the frequency, for the various modes of this composite system. This problem of energy sharing may in fact be regarded as that of VA diagnostics; namely, of identifying the vibration transmission path from measured responses (see Chapter 6).

The results of travelling-wave response measurements may be used generally for estimating the properties of the media. Moreover, reflections or 'echoes' from various obstacles may be used for the detection of various defects or cracks in materials and/or structures. Such diagnostics are usually made in ultrasonic tests, where signals of very high frequency(ies) are applied to the specimen, or structural element. These ultrasonic inspection methods have their limitations, however, one of which is their general inapplicability to operating machines or structures. Thus, we consider in this book mainly those methods that are based on an analysis of 'natural' inservice waves, with predominant frequencies in the audible range (up to about 20000 Hz).

Consider first the identification problem of estimating the propagation velocity from measured travelling-wave responses. We start with the simplest case of a single-mode free nondispersive wave, such as that in the above example of a stretched wire. Assume that the wave propagates 'from left to right', and that the system's responses u(x,t) are measured at two points 1 and 2 with coordinates x_1 and $x_2 = x_1 + l$, respectively. The response characteristics in general may be system-dependent: it may be the lateral deflection of the wire, or the acoustic pressure within a pipe, as measured by a microphone or hydrophone, or the acceleration of a rod in the case of axial waves etc. In any case, the basic property of a nondispersive wave, without attenuation, implies that, if a certain signal $u_1(t)$ is registered at point 1, then a signal with the same undistorted shape will be registered at point 2, after the time delay:

$$\theta = \frac{l}{c} ,$$

or

$$u_2(t) = u_1(t-\theta) ,$$

$$\text{where} \quad u_2(t) = u(x_1 + l, t) , \quad u_1(t) = u(x_1,t) \tag{4.1}$$

Here, θ is the transit time for the wave to propagate along the distance l between the sensors at points 1 and 2. Thus, the identification algorithm is reduced to a simple estimation of the time delay between two (nominally identical) traces, $u_1(t)$ and $u_2(t)$. This can be done for pulse-like signals, even in the presence of some noise background or measurement errors, if the time shift between the peaks of the pulse-like $u_1(t)$ and $u_2(t)$ is estimated. Moreover, if these peaks are found to be of unequal height, owing to propagation losses, the latter may be directly estimated from the ratio of these peak values.

Consider now a somewhat more complex case of a sustained wave, excited by a certain time-variant loading, applied somewhere to the left of point 1. For definiteness, we assume at first that this loading is a stationary random process. In this case, a direct identification of the reproduced or corresponding points of signals 1 and 2 may prove to be rather difficult, especially in the presence of measurement noise and/or errors. In this case, a correlational analysis of the response signals may be used with advantage. Multiplying both sides of eqn. (4.1), written in the form $u_2(t + \tau) = u_1(t + \tau - \theta)$, by $u_1(t)$ and applying the averaging operation yields.

$$K_{12}(\tau) = K_{11}(\tau - \theta)$$

$$K_{11}(\tau) = \langle u_1(t)\, u_1(t + \tau)\rangle \quad , \quad K_{12}(\tau) = \langle u_1(t)\, u_2(t + \tau)\rangle \tag{4.2}$$

Thus, the crosscorrelation function of the responses at two points reproduces the response autocorrelation function at the first point, with a time delay $\theta = l/c$. The latter, once again, is equal to the transit time of the wave propagation from point 1 to point 2. However, one of the basic properties of an autocorrelation function is that its absolute maximum is at zero time lag (see Chapter 1). Therefore, the crosscorrelation function $K_{12}(\tau)$ reaches its absolute maximum at $\tau = \tau_m = l/c$, so that the propagation velocity may be estimated easily, from the measured responses at two points, as $c = l/\tau_m$. Moreover, the normalised maximal crosscorrelation factor:

$$\rho_{12} = \frac{K_{12max}(\tau)}{(K_{11}(0)\, K_{22}(0))^{1/2}} \quad , \quad K_{12max}(\tau) = K_{12}(\tau_m) \tag{4.3}$$

provides a direct estimate of the propagation losses or the attenuation factor of the waves. The basic advantage of this crosscorrelational approach is that it permits one to filter out measurement noises and/or errors, which are generally mutually uncorrelated and also uncorrelated with the propagating waves. Therefore, the crosscorrelation function $K_{12}(\tau)$ is noisefree (and therefore may still be used to estimate τ_m and c), although $K_{11}(\tau)$ and $K_{22}(\tau)$ in general are not.

It is quite easy, using this distributed system, with finite transit times or delays, to estimate the direction of the wave propagation. Indeed, suppose that the latter is not known in advance, and therefore the selection of u_1 and u_2 (that is, of the response signal which should be taken with a time delay to estimate the crosscorrelation function) is made in a random fashion. The 'wrong' choice will be reflected then automatically in that τ_m will be negative; i.e. the crosscorrelation function will be shifted towards the negative rather than the positive semiaxis.

The above crosscorrelational approach is advantageous in general for arbitrary waveforms. It is used in the testing of large structures with various test signals, such as sweep sine excitation [52] which leads to a pulse-like wave (these topics are considered in more detail in Chapters 5 and 6). The 'natural' waves in certain cases may also be periodic, or almost periodic, rather than random. Such is the case, for example, with a pulse wave in a human artery. Its propagation velocity has been measured *in vivo* by the crosscorrelational method in [15], the 'natural' signal being periodic but not sinusoidal. Such measurements provide an index of the elasticity of the artery, which is quite important for the diagnosis of arteriosclerosis and aneurysm [15].

Consider now the possibility of extending this approach to dispersive waves, such as one-dimensional flexural waves in thin beams or plates. Quite a natural approach to such an extension is as follows. The analyses of both the response signals, $u_1(t)$ and $u_2(t)$, are made separately in various narrow frequency bands with the use of two sets of bandpass filters. Within each band with an almost constant value of propagation velocity $c(\omega)$ (where ω is now the central frequency of the band), the above crosscorrelational analysis can be performed. This seems like an attempt 'to deceive Nature' by pretending that the other frequency components of the responses do not exist. The approach does indeed work, provided that the results are interpreted properly; however, it is found in the end that Nature cannot be deceived. Specifically, a rigorous mathematical analysis shows [39] that, for random-in-time waves, the propagation velocity may still be estimated as $c = l/\tau_m$; however, it is not $c(\omega)$ itself, but rather a so-called group velocity $c_g(\omega) = c/[1 - (\omega/c)(dc/d\omega)]$. This should be expected, since the latter is the 'apparent' or 'effective' propagation velocity of a wave packet with close frequencies of the component waves. In the long run, the results of separate analyses in various frequency bands are seen to be coupled if we try to reconstruct the original curve $c(\omega)$ from these response data!

In general, the group velocity may be significantly different from the phase velocity; for example, for flexural waves $c_g = 2c$. The response cross-spectrum can, in principle, provide an estimate of the phase-velocity value at a certain particular frequency only if this cross-spectral density is estimated within a zero bandwidth. As seen in Chapter 2, in view of the Uncertainty Principle, such an estimate from a finite-length sample of a random process is always found to be 'smeared' within a finite (albeit small) bandwidth. Therefore, only the group velocity may be estimated by processing random response data. This is a case where the Uncertainty Principle (see Chapter 2) may lead to quite drastic changes in estimates, compared with those corresponding to 'ideal' input data, for media with a strong dispersion. On the other hand, with periodic-in-time waves, such as pulse waves in a human artery, this approach can provide estimates of the phase rather than the group velocity.

Furthermore, in the above approach the values of the maximal normalised crosscorrelation function ρ_{12} (eqn. (4.3)) are less than unity, even in the absence of any propagation losses and measurement noise. The reason is the effect of the spatial spread of waves with slightly different frequency components, within a finite bandwidth. This reduction in the crosscorrelation has been calculated in [39]. The resulting correction factors should be applied whenever attempts are made to estimate propagation losses and/or reflection factors (as considered later in this Chapter) from response data.

This procedure can be extended in principle to the case of multiple transmission paths between points 1 and 2 with different transit times (delays) $\theta_i = l_i/c_i$. In this case, the crosscorrelation curve will contain several 'pulses' with identical shape, each of them corresponding to a certain path. The identification procedure is especially easy when the differences between the various θ_i are much higher than the correlation time of the wave, so that these pulses do not overlap; note that this requirement may preclude the use of a too small filter bandwidth, in the case of dispersive waves. The values of ρ at a certain θ_i may then be used to estimate which part of the energy of the total wave is transmitted through the corresponding path. This problem is considered in more detail in Chapter 6.

Assume now that some restraint is placed at a certain point of a one-dimensional system, such as an additional mass attached to a stretched string, or a stiffener or an electronic block attached to the skin of an aircraft or rocket structure. When a wave reaches this restrained point, part of this incident wave may be transmitted, and it will propagate further in the same direction. The other part, or reflected wave, will propagate backwards. This split of the incident wave may be described quantitatively by a reflection factor. This is the ratio of the responses (say, the pressures in acoustic waves or the lateral deflections in bending waves in beams) in the reflected and incident waves. The crosscorrelational approach may sometimes be used to detect the reflected wave and possibly to estimate the reflection factor from the measured response data. This may be of interest in some applications, such as the inservice detection of damage or cracks in structures. Whenever no special high-frequency ultrasonic signals can be applied, one should rely on normal operational vibrations with wavelengths high compared not only with the crack length but also with the thickness of the structural element. For this application, a partial reflection of a wave would imply crack initiation, whereas the reflection factor would be a direct measure of the crack size or depth (a more detailed analysis of crack-detection problems is presented in Chapter 9).

It should be stressed that, in general, the conditions at various reflective boundaries may be applied to various characteristics of the response in a given system. For example, at the end of a fluid column within a pipe, the condition of zero pressure (open end) or zero velocity (closed end) may be given, or, more generally, the values of the pressure and velocity at the end may be related in a particular way. One of the ends of a bar with a propagating axial stress wave may be clamped (a boundary condition of zero displacement) or free (a boundary condition of zero stress). By the way, in the latter case a stress wave changes its sign after reflection, since the total stress at the end should be zero, so that a compressive stress wave is reflected as a tensile one; this may be of importance for brittle materials, which can often bear quite high compressive stresses but fracture under relatively small tensile stresses.

Alternatively, if a single response variable u(x,t) is introduced, the boundary conditions may include both this variable and also its spatial derivatives, so that a single reflection factor will, in general, be complex. Therefore, the following model of a single real reflection factor, which accounts for the change in amplitude, but not in the phase, of the reflected wave is rather limited. It is presented here mostly for illustrative purposes and for qualitative rather than quantitative diagnostics (simple detection of a reflected wave).

Figure 4.1 illustrates the scheme of the system. The total response at any

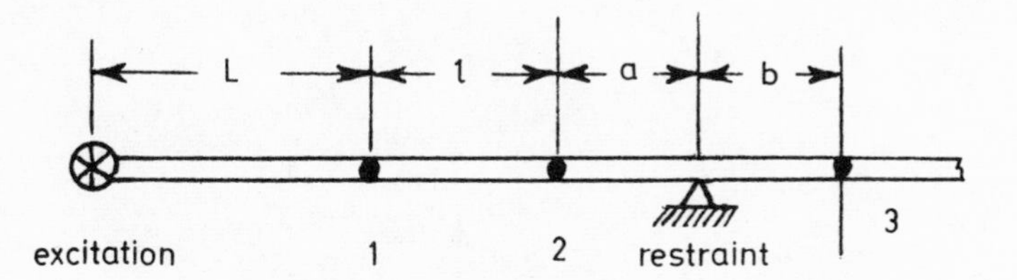

Fig. 4.1 Sketch of a system for which the reflection and/or transmission factors are estimated from the measured responses at points 1, 2 and/or 3

point may be represented as a sum of the responses due to the incident and reflected waves. Therefore, for negligible propagation losses:

$$u_1 = u(t - \frac{L}{c_g}) + k\,u(t - \frac{L + 2l + 2a}{c_g})$$

$$u_2 = u(t - \frac{L + l}{c_g}) + k\,u(t - \frac{L + l + 2a}{c_g}) \tag{4.4}$$

where the first terms on the right-hand-sides correspond to the incident waves and the second ones — proportional to the reflection factor k (real!) — to the reflected waves; c_g is the group velocity. It is evident now that the crosscorrelation function $K_{12}(\tau) = \langle u_1(t)\, u_2(t + \tau)\rangle$ should contain four terms with the corresponding time shifts at their absolute maxima being $\tau_m = l/c_g$, $\tau_m = -l/c_g$, $\tau_m = (l + 2a)/c_g$, $\tau_m = -(l + 2a)/c_g$. For a sufficiently small correlation time of the random wave signal, these 'pulses' of $K_{12}(\tau)$ should not overlap. Then the reflection factor may be estimated as the ratio of the maxima at, say, $\tau_m = (l + 2a)/c_g$ and $\tau_m = l/c_g$ (with an appropriate correction factor for the loss of coherency of the wave packets in dispersive waves). The apparent overlap of the various neighbouring components of the crosscorrelation function may be reduced sometimes by a nonlinear (logarithmic) transformation, as described in Chapter 6 (the cepstral analysis).

As explained above, the reflection conditions in many systems may be more complex, so that much more sophisticated algorithms of response data processing may be required for their quantitative identification. The above simple case illustrates that, in the presence of a wave reflection, certain additional 'pulses' of crosscorrelation function will appear at 'negative' time shifts. They can be used, say, for the detection of defects or cracks. Similarly, the crosscorrelation function $K_{23}(\tau)$ of the responses at points 2 and 3, situated at both sides of the 'obstacle' or point with a restraint (see Fig. 4.1),

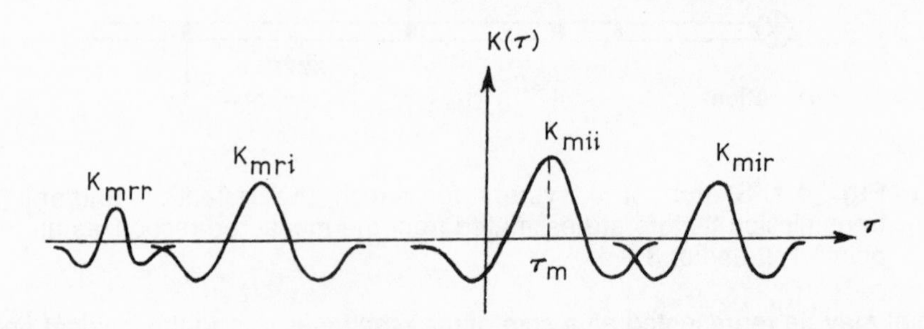

Fig. 4.2 Crosscorrelation function $K_{12}(\tau) = \langle u_1(t)\, u_2(t+\tau)\rangle$ between responses at points 1 and 2 of the system in Fig. 4.1

should have two 'pulses' with $\tau_m = (a + b)/c_g$ and $\tau_m = (b - 2a)/c_g$, the latter one corresponding to the crosscorrelation between the transmitted and reflected waves. Once again, the identification of such secondary pulses is easiest when the pulses do not overlap; this requirement may impose the restriction of not too small a frequency bandwidth for the analysis, for waves with dispersion (the response correlation time should not be too high).

In relation to crack detection, an important phenomenon of diffraction should be mentioned here, basically two- or three-dimensional. Consider a plane wave in a two-dimensional medium arriving at a certain obstacle. Whenever the characteristic dimension of the latter is high compared with the wavelength, a complete reflection becomes possible, with the formation of a distinct shadow zone behind the obstacle. However, waves with wavelengths of the same order as the dimensions of the obstacle may partly circumvent it. Such an effect can be seen in harbours where sea waves may easily circumvent a lonely single pile but not a long breakwater dam. Another example is that of two persons walking in a thick forest. When they become separated by more than, say, two dozen metres, it may become impossible for them to see each other. But they may hear each other at much higher distances. The reason is that, while trees and bushes, i.e. typical obstacles in the forest, are extremely large compared with the wavelengths of light (which are much less than one millimetre), they are of the same order as, and possibly smaller than, the sound wavelengths. Therefore, the forest is apparently somewhat 'transparent' to audible sounds because of the diffraction phenomenon.

It is this phenomenon that makes ultrasonic diagnostics particularly suitable for crack detection and identification, since very low wavelengths in the signals may be used in these techniques. This means that by reducing the signal wavelength one can, in general, improve the resolution, so that the smaller cracks can be detected. On the other hand, with 'natural' inservice signals in the audible frequency range, the resolution may be poor. To improve it , quite sophisticated data-processing algorithms may be required.

Concluding this chapter, we must mention briefly an important topic: nonlinear waves. There exist a large variety of nonlinear-wave phenomena, which are quite complex for both mathematical analysis and physical interpretation. In stress waves in solids, the nonlinearities are usually due to nonlinear stress-strain relations (physical nonlinearity). In flexural elastic waves in thin-walled structures, they may also be due to nonlinear relations between the strains and the lateral deflections of the structure (geometrical nonlinearity for not-too-small deflections).

Any extensive analysis of nonlinear waves is beyond the scope of this book, and only certain basic simple facts are mentioned here. Nonlinearity, say, of axial stress waves in a bar with a nonlinear governing stress-strain curve of the material implies that the propagation velocity is now state-dependent rather than a constant. This implies that wave signals in general may become distorted during propagation, even in a nondispersive medium. In particular, if a sinusoidal timewise excitation is applied at one end of a waveguide, the higher harmonics of this signal may appear along the path. In certain media, these harmonics may become amplified to such an extent that a shock wave is formed. Thus, whenever a waveform (say, of an acoustic signal in a gas) is transformed from a sinusoidal to a sawtooth one, owing to these higher harmonics, the smooth wave becomes unstable. Then a shock wave is formed, similar to that in a sonic boom from a supersonic airliner, with drastic changes or jumps in the state variables across the shock.

From the point of view of acoustical diagnostics, nonlinear phenomena may be of interest as sources of additional information on the media properties. For example, in the above case of a sinusoidal input signal, this information may be provided by the higher harmonics of the input frequency. Certain algorithms of diagnostics, based on response measurements of nonlinear waves, may be found in [38]. Whenever a random input signal is applied at the end of a waveguide, the nonlinearities of the media properties may lead to a nonzero crosscorrelation between the signal's components in different frequency bands. This effect may be regarded as the random-case counterpart of the above effect of higher-harmonics generation by nonlinearities in the medium. A similar effect is observed in nonlinear oscillations in lumped-parameter systems — see Chapter 9, where its application to crack detection is described.

Chapter 5 Dynamic Characteristics of Linear Systems

This chapter starts with a previously-promised general definition [54]. A dynamic system is called linear if the superposition principle holds: i.e. the system's response to a weighted sum of arbitrary input signals equals the weighted sum of the responses to these signals, with the same weighting factors. By input signals we mean here both the external forces (in general, time-variant) and the initial disturbances. Furthermore, these input signals may be applied to any mass(es) of a system with lumped parameters or at any point of a structure with distributed parameters. The response(s) of a mechanical system or structure may be its displacement, velocity, acceleration, stress etc. at any point(s).

It is clear, in view of this definition, that a linear system may be described completely by its responses to a certain set of baseline, or reference input, signals, provided that any given input signal may be considered to be composed of these baseline signals. This can be done either in the time domain or in the frequency domain. In the first case, the impulse response function $G(t,t')$ is used, which yields the system's response at a time instant t to a Dirac delta-function external input $\delta(t-t')$ applied at the time instant $t' < t$. In a multimass, or multi-degree-of-freedom (MDOF), system with lumped parameters, this function, as well as its frequency-domain counterpart, or transfer function, is generally used with two subscripts. The first one of these corresponds to the responding mass or degree of freedom (DOF), the second one to the excited mass. In a structure with distributed parameters, this function also depends on two spatial positions — namely, those of the responding point $\underset{\sim}{r}$ and the excited point $\underset{\sim}{r}'$. In mathematical physics, such a function $G(\underset{\sim}{r},t,\underset{\sim}{r}',t')$ is usually called a Green's function.

An important property of any impulse response function of common mechanical systems is that of causality: i.e. $G(t,t') = 0$ for $t < t'$. It implies that the motion cannot start until the excitation is applied. It is interesting that no similar condition exists for the Green's function dependence on spatial coordinates, which, in general, may be 'two-sided'. For example, the weight of a person, standing on a footbridge, is felt at both ends of the bridge; the latter, in fact, is completely unaware of such concepts as 'past' and 'future', at least as far as static loading is concerned: both spatial directions are alike to it. Certain systems with feedback control may sometimes be noncausal. For example, the active vibration control of a vehicle, obtained by applying a control force to its rear suspension, need not necessarily be causal if it is based on a control signal as provided by a vibration sensor at the front suspension; the latter may then provide some 'prediction' of the track irregularities to be faced by the rear wheels.

The form of representation of the response, x(t), in terms of the input signal and the impulse response function can be established easily. Indeed, by introducing a small time step $\Delta t = (t - t_0)/N$ (t_0 is an initial time instant), any input signal f(t) can be approximated by a sequence of pulses, where the ith pulse has the strength or intensity $f(t_i)\,\Delta t$ and provides a contribution $G(t,t_i)\,f(t_i)\,\Delta t$ to the overall response, x(t). The latter, in view of the superposition principle, may be represented as a sum of these contributions, which in the limiting case $\Delta t \to 0$ leads to the familiar convolution integral (see Chapter 2):

$$x(t) = \lim_{\Delta t \to 0} \sum_{i=1}^{N} G(t,t_i)\, f(t_i)\, \Delta t = \int_{t_0}^{t} G(t,t')\, f(t')\, dt' \qquad (5.1)$$

In this chapter, time-invariant systems are mainly considered. For such a system, any one of its impulse response functions should depend only on a single variable, namely the time shift $\theta = t - t'$, rather than on both t and t′. An alternative description of such systems is provided in the frequency domain by the transfer function. The latter is in fact the Fourier transform (FT) of $G(\theta)$. It may be defined alternatively as the complex amplitude of a steady-state system's response to the input signal $\exp(i\,\omega\, t)$; such a response signal is represented as $H(\omega)\exp(i\,\omega\, t)$.

Some response-to-input ratios (where both these signals are proportional to $\exp(i\,\omega\, t)$) have specific titles in mechanics and acoustics. For example, the ratio of the displacement to the force is called the dynamic compliance, or flexibility (with its reciprocal called the dynamic stiffness), whereas the ratio of velocity to force is called the mobility (with its reciprocal called the mechanical impedance). Other names for various transfer functions can be found in [22,34].

The convolution operation (5.1), after Fourier transformation into the frequency domain, reduces to a simple multiplication:

$$X(\omega) = H(\omega)\, F(\omega) \tag{5.2}$$

where X and F are the FTs of x(t) and f(t), respectively.

Since $H(\omega)$ is a complex function, various methods of presentation are possible. Two of the most common are as follows:

(i) The amplitude and phase frequency response functions, AFR and PFR, i.e. the modulus $A(\omega) = |H(\omega)|$ and the argument, or phase, $\varphi(\omega) = \arg H(\omega)$, of the transfer function, respectively. In Fig. 5.1(a), these functions are illustrated

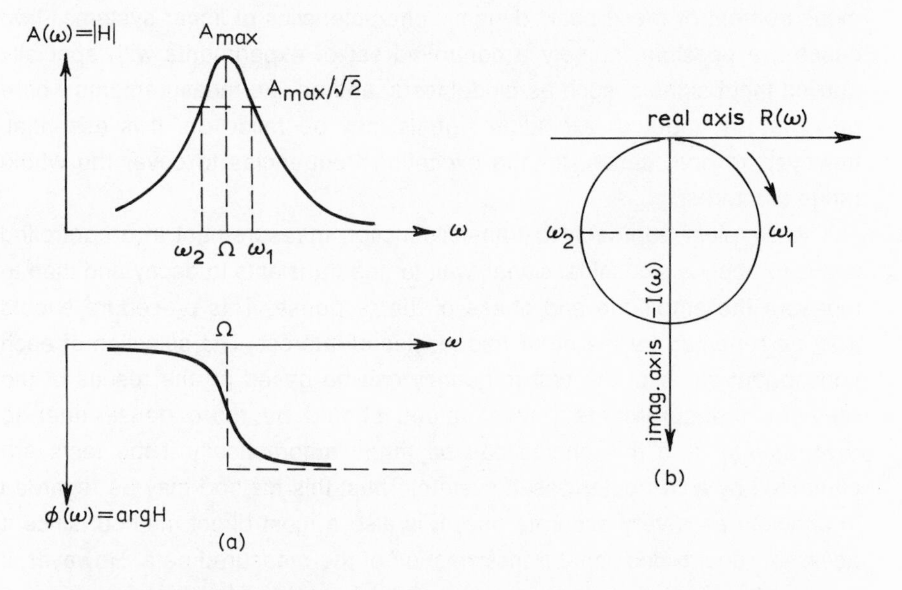

Fig. 5.1 Alternative representations of the transfer function $H(\omega)$ of a damped SDOF system: frequency responses (a) and vector diagram or Nyquist plot (b)

for a lightly damped SDOF system. At the system's natural frequency, a peak of amplitude $A(\omega) = |H(\omega)|$, and a drastic variation in phase, are observed; with decreasing damping, the former becomes narrower, and the latter steeper.

(ii) A vector diagram, or Nyquist plot, provides a locus of $H(\omega)$ for various ω in the complex H-plane. In fact, it is a graph of the real part $R(\omega)$ versus the imaginary part $I(\omega)$ of $H(\omega)$, with the frequency ω entering this plot only implicitly (as a parameter). Such a plot, for a lightly damped SDOF system, is presented in Fig. 5.1(b). It may be regarded as an amplitude-phase curve as well, if polar coordinates are introduced in the complex H-plane; i.e. $|H(\omega)| = [R^2(\omega) + I^2(\omega)]^{1/2}$, $\varphi(\omega) = \arg H(\omega) = \arctan [I(\omega)/R(\omega)]$. The parameter ω increases along the curve in a clockwise direction. It may be noted that, while major variations in the amplitude and phase responses of a lightly damped system correspond to a narrow frequency range, in the vicinity of the natural frequency its Nyquist plot is 'stretched' along the frequency axis; i.e. the above range corresponds to the biggest part of this curve. Therefore, the Nyquist plot may be more suitable for estimating the system's parameters from measured transfer-function data.

Consider now various types of excitation that may be used for the measurement of these basic dynamic characteristics of linear systems. Two cases are possible, namely a controlled set of experiments with specially applied input signals, such as modal tests, and on-line measurements where only 'natural' input or excitation signals may be relied on. It is essential, however, in both cases, for the excitation frequencies to cover the whole range of interest.

The simplest approach to transfer-function measurement in a controlled test is to apply a sinusoidal signal, wait for the transients to decay and then to measure the amplitude and phase of the response. This procedure should then be repeated for the other frequencies of interest. The selection of each subsequent value of the test frequency can be based on the results of the previous measurements (these values should be more dense near to resonances), and this choice can be made automatically if the tests are controlled by a computer-based system. Thus, this method may be regarded in principle as a very accurate one. It is also a most direct method, since it does not require additional transformation of the measured data. However, it is also the most time-consuming method because of the slowness of a sequential survey of the required frequency range.

A natural approach to overcoming this shortcoming is to scan the prescribed frequency range with as high a sine frequency sweep rate as possible, without distorting the results. A check on the results can be made by making the measurements twice — first with an increasing excitation frequency and then with a decreasing excitation frequency. If the results are the same, and particularly if every peak of the AFR is attained at the same value of the excitation frequency*, then the sweep rate in this sine sweep test may be regarded as sufficiently low for the direct use of measured amplitudes and phases. With a higher sweep rate, the excitation should be regarded as transient, and the reconstruction of the desired characteristics, say AFR and PFR, requires a special data-processing procedure. The required relations for a linear sweep (constant sweep rate) are presented in [52].

Another type of excitation, which permits a quick survey of the whole frequency range of interest, is broadband random excitation. In this case, in fact, many frequency components of the response are excited simultaneously. This type of excitation is used both in special tests, where it may be produced by various commercial shakers, say, for fatigue testing, and in on-line measurements. In the latter case, the excitation, or input signal, cannot usually be controlled, but, if possible, should be measured carefully, together with the response signal. Such measurements are used, for example, to identify the dynamic characteristics of a car from response data obtained during a ride along a rough road.

The desired transfer function(s) $H(\omega)$ may be obtained in two ways. Let the Fourier-transformed input signal F and the response X be related by $H(\omega)$ as $X = H F$; then, multiplying this relation in turn by F and X, we obtain the two relations:

$$\Phi_{fx}(\omega) = H(\omega)\, \Phi_{ff}(\omega) \quad , \quad \Phi_{xx}(\omega) = H(\omega)\, \Phi_{fx}(\omega) \tag{5.3}$$

where $\Phi_{ff}(\omega)$ and $\Phi_{xx}(\omega)$ are the spectral densities of f(t) and x(t), respectively, whereas $\Phi_{fx}(\omega)$ is the cross-spectral density of the input and response signals.

Therefore, two estimates of H from measured input/output data are possible:

$$H_1(\omega) = \frac{\Phi_{fx}(\omega)}{\Phi_{ff}(\omega)} \quad \text{and} \quad H_2(\omega) = \frac{\Phi_{xx}(\omega)}{\Phi_{fx}(\omega)} \tag{5.4}$$

With perfect measurement, these estimates should be identical, as can be seen from the following formula for the coherence function γ^2, as obtained

---

* During fast passage through resonance of the excitation frequency (an important phenomenon for rotating shafts), the peak amplitude of the response is attained with some time delay — i.e. at a somewhat higher-than-resonant frequency in an 'acceleration' test and at a lower frequency in a 'deceleration' test.

from eqns. (5.4) (see the definition in Chapter 1):

$$\gamma^2(\omega) = \frac{H_1(\omega)}{H_2(\omega)} \tag{5.5}$$

Indeed, if neither of the two measured signals is corrupted by noise, then $\gamma^2 = 1$. Moreover, this condition, $\gamma^2 = 1$, can be used as a check on the accuracy of the estimates. In real applications, it may be found, however, that the coherence is less than unity ($H_1 < H_2$), owing to noise and/or measurement errors. In such cases, H_2 may be a more accurate estimate, near to the resonances, than H_1, while the reverse is true at the antiresonances (the minima of the response spectrum) [12]. The accuracy of estimates of H_1 and H_2 using different sample lengths of the measured signals has been studied in [12,22].

In a white-noise input excitation, a crosscorrelational analysis of the input and response signals provides a direct estimate of the system's impulse-response function. In fact, if in the first eqn. (5.3) $\Phi_{ff}(\omega)$ is constant, the inverse Fourier transform of this equality yields $K_{fx}(\tau) = G(\tau)$. Moreover, this estimate provides a simple check, or 'self-control' of the basic white-noise assumption, since the condition $K_{fx}(\tau) = 0$ should be satisfied for every $\tau < 0$ in view of the previously mentioned causality property of $G(\theta)$.

In this connection, the following interesting problem may be considered. Suppose that it is not known in advance which of the two available measured signals should be regarded as the 'input' and which as the 'output' or the 'response' of a linear causal dynamic system. Can the identity of these signals be established by some appropriate processing of their samples? This question may not be of only academic interest if the given system is regarded as a real 'black box'; i.e. if its physics is completely unknown. Moreover, such a problem may also arise for complex mechanical systems where both the signals are measured responses at two different points; in this case, in fact, the direction of the random signal propagation within the system is sought.

In view of the above property $K_{fx}(\tau) = G(\tau)$ for a white-noise f(t), one may try the following algorithm. A crosscorrelation function $K_{fx}(\tau)$ of the given signals is calculated with an arbitrary selection of input and response signals. For a proper choice (with positive time delays τ assigned to the response), $K_{fx}(\tau)$ is zero for $\tau < 0$; otherwise, it is zero for $\tau > 0$, if the input signal is indeed a white noise. This extreme case, where $K_{fx}(\tau) = G(\tau)$, is illustrated in Fig. 5.2 by a dashed line. In the opposite extreme case of a slow or quasistatic response, the above problem cannot be solved at all. Indeed, in this case of a static

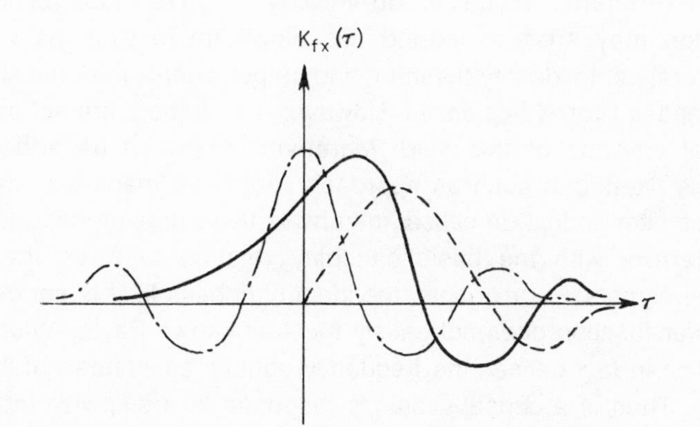

Fig. 5.2 Crosscorrelation functions between the input and response signals of a linear system: general case (full line), white-noise-input case (dashed line) and quasistatic case (dash-dot line)

rather than a dynamic linear coupling between f(t) and x(t), we have a simple proportionality relation x(t) = c f(t). Therefore, the crosscorrelation function $K_{fx}(\tau)$ simply reproduces, up to a constant factor, the input autocorrelation function $K_{ff}(\tau)$ and thus is an even function of τ. This symmetric case, with both input and response signals being 'of equal standing', is illustrated in Fig. 5.2 by the dash-dotted line.

The solid line in Fig. 5.2 corresponds to the typical general case, where the correlation time of the input signal may be of the same order as the characteristic time constant of the system. It illustrates some asymmetric 'shift' of the crosscorrelation function $K_{fx}(\tau)$ towards the positive semiaxis; i.e. in the direction of the chosen response signal. This case is an intermediate one between the above extreme cases of a 'one-sided' $K_{fx}(\tau) = G(\tau)$ and of a symmetric $K_{fx}(\tau) = c\,K_{ff}(\tau)$. However, it is not easy to derive rigorously which specific feature of $K_{fx}(\tau)$ (or, as a mathematician would say, what functional of $K_{fx}(\tau)$) describes this asymmetry in the general case and thus may be used as a universal criterion to identify properly the input and response signals.

It was shown in Chapter 4 that the desired feature can be identified easily for a system with a pure time delay; in this case, it is the point τ_m of the absolute maximum of $K_{fx}(\tau)$. This case corresponds, in particular, to a nondispersive wave propagation in the system with distributed parameters, with a finite transit time between the two points at which the response signals are measured. The case of dispersive wave propagation may also be treated, as shown in Chapter 4, by detecting the narrowband components from both signals; the corresponding time shift τ_m at the absolute maximum of

$K_{fx}(\tau)$ then determines the group velocity of wave propagation. A strong temptation may arise to extend this algorithm to systems with lumped parameters; i.e. to identify the input and output signals from the sign of τ_m for the bandpass filtered responses. However, the authors are not aware of any universal criterion of this kind. Moreover, it should be added that any bandpass filtering in such an approach should be made with care; i.e. the bandpass filter should be causal, otherwise this signal-processing procedure may interfere with the basic causality relation between the input and response signals. For example, the ideal bandpass filter is not causal, since its transfer function does not satisfy the well known Payley-Wiener criterion [54], which in fact defines the frequency-domain counterpart of the causality property. Thus, if a causal system's response to a stepwise input signal is passed through an ideal narrowbandpass filter, the inverse Fourier transform of this narrowband filter output may be nonzero even at the time instants preceding the original input step! This clearly shows that the causality relation between the measured signals may be an artefact of their processing procedure if the latter involves noncausal filtering.

Returning to the basic problem of estimating $H(\omega)$, we may mention yet another case of on-line measurement. Often, only the system's response can be measured throughout its routine proof, or inservice, tests, whereas the input or excitation signal may be inaccessible to measurement — either because of practical difficulties, or because its origin may not be clear enough, in some cases. For example, many vibration sensors and strain gauges may be used to monitor the piping vibrations of a nuclear power plant during its hydraulic tests and/or inservice life; however, measurements of fluid-pressure oscillations within the pipes (input signals) may be possible, if at all, only on a much smaller scale. With such 'excess' available response data, a natural question arises: how one can use it to get some information on the machine or structural system? The above example is rather typical in that similar situations may arise for many structures interacting with a fluid flow.

This problem may be solved, at least partly, if some reasonable assumption is made concerning the shape of the input signals' spectra. When these signals are mainly due to pressure oscillations in a turbulent boundary layer on a structure's surface, interacting with a fluid flow, these spectra may often be regarded as being broadband compared with the bandwidths of the resonant peaks of the amplitude-frequency-response (AFR) function. Since any such bandwidth is of the order of the corresponding structure's modal damping ratio, which is usually relatively small, this assumption may be regarded as plausible, at least for unseparated unconfined flows, and it is used frequently in the next chapter. (In a piping system, the flow is not completely unconfined, and narrowband excitation due to acoustic resonances within the pipes may also be present. This possibility is ignored at this stage, assuming that the acoustical resonances are sufficiently separated from the structural ones, so that the latter may be studied independently.)

Acceptance of this basic broadband excitation hypothesis provides the possibility of estimating the system's AFR directly from the response spectral density $\Phi_{xx}(\omega)$. Indeed, since the input spectral density may be approximately assumed to be constant in the vicinity of any given i-th resonant peak, $|H^2(\omega)|$ is proportional to $\Phi_{xx}(\omega)$ in this frequency range. The unknown proportionality factors c_i may, in general, be different for different spectral peaks; however, they may be of secondary importance. In any case, this approach may at least provide reasonable estimates of the system's natural frequencies and modal damping ratios. It should be noted, though, that these estimates may be highly sensitive to measurement noise and errors; furthermore, the requirements concerning the sample length of the recorded signals are much more stringent compared with the cases where both input and output signals are measured and cross-spectral densities (or crosscorrelation functions) are used for identification*. Moreover, all the information on the phase frequency responses (PFR) is lost within this approach, and therefore it may be inadequate for those diagnostic purposes that require such information.

Tests with transient excitation should now be mentioned. For many structural components, impulse-type excitation, from a hammer blow, can be implemented most easily, since it does not require any special shakers. Moreover, the response in such a test provides directly an estimate of the impulse response $G(\tau)$ in view of the definition of the latter; the transfer function may then be calculated, if desired, as the Fourier transform of $G(\tau)$. The impacts due to various 'natural' sources are also used readily by engineers, whenever appropriate measurements are possible, to identify the basic dynamic properties of complex structural systems. For example, a drastic change in the coolant flow during hydraulic tests of the piping system of a nuclear power plant may lead to a mild 'waterhammer' phenomenon, which provides the required excitation of all the pipes. Other examples are those of a car, striking a 'bump' on an otherwise smooth road, and of a tall building shaking during an earthquake. In such cases, it is advisable to make the necessary measurements of the corresponding transient responses.

It seems relevant now to discuss the problem of estimating a system's

---

* In the latter cases, the sample length is not of importance in the ideal situation of absolutely accurate measurements, since the lack of a true statistical representation of the random input and response is of minor importance, as long as both of them are measured; of course, when it is necessary to filter out measurement noises, it is desirable to increase the available sample length.

parameters from its measured transfer functions, or impulse response functions. Such an estimation may be of importance in cases where only a few key parameters are used as indexes for diagnostic purposes. For example, monitoring of the inservice variations of the natural frequency(ies) of certain given mode(s) may provide information on the deterioration of a machine component (see Chapter 9); the stability margin of a structure may be estimated from its minimal modal damping ratio, or the damping ratio of a certain given mode (Chapter 7). The accuracy requirements of parameter estimates may be quite stringent, especially when their variations are used for diagnostics (as in the first example).

The problem of parameter estimation may be reduced ultimately to that of a curve-fitting, and many suitable least-squares-fit computer routines are available. This reduction, however, should always be made in some 'optimal' way, using algorithms that are the least sensitive to the various errors in given data. In particular, the necessity for the estimation procedure to be 'robust', with respect to any small inadequacies in the basic mathematical model, should be mentioned here. For example, even a small nonlinearity in the restoring force of an SDOF system with broadband random excitation may lead to a quite significant broadening in its response spectrum, compared with the 'natural' damping-controlled system bandwidth, so that a direct estimate of the system's small damping, from this spectrum, may be erroneous.

We mainly consider here lightly damped systems with distinct resonant peaks, rather than systems with a high modal overlap ratio, which have smooth frequency responses. However, the case of two modes with close natural frequencies is also discussed.

The simplest case of parameter estimation is for the decaying free oscillations in an SDOF system excited by some initial disturbance or impact. These oscillations are described by the product of a decaying exponent and a sinusoid. The frequency of the latter provides an estimate of the system's damped natural frequency, whereas the damping ratio may be estimated as:

$$\frac{\alpha}{\Omega} = (2\pi)^{-1} \ln \left(\frac{A_n}{A_{n+1}}\right) \approx (2\pi)^{-1} \left(\frac{\Delta A_n}{A_n}\right) \tag{5.6}$$

Here, A_n and A_{n+1} are two successive (nth and n+1th) maximal values (amplitudes) of the response, and the last expression is valid approximately for a lightly damped system ($\alpha/\Omega << 1$), where $\Delta A_n = A_n - A_{n+1}$ is much less than A_n.

In an MDOF system, the estimation of modal parameters from impulse response(s) G(t), obtained in an impact test, implies an approximation of G(t) by a sum of terms of the form $C_i \exp(-\alpha_i t) \sin(\Omega_i t + \varphi_i)$, where Ω_i, α_i are the ith natural frequency and the ith modal damping factor, respectively, whereas C_i and φ_i are the amplitude and phase, respectively, of the ith modal response. Various procedures for such an approximation may be found in [11], and certain difficulties, arising for closely spaced Ω_i, should be mentioned here. The proper choice of the number of terms in the approximating sum for G(t) (i.e. the 'effective' number of degrees of freedom, or DOF) is of extreme importance in any of these procedures. Quite sophisticated rules for this choice do exist, which may be used if this number of DOF is not known in advance [2]. However, for lightly damped mechanical systems, the following simple rule often works well: the number of effective DOF may be equated to the number of peaks in the system's AFR. Of course, this criterion may only work well generally if all these DOF are represented adequately in the measured response signal(s).

Turning now to estimation in the frequency domain, we start once again with an SDOF system. The simplest estimates, as obtained from the AFR curve (see Fig. 5.1(a)), are as follows: Ω is the maximum of $A(\omega) = |H(\omega)|$, whereas the damping ratio is:

$$\frac{\alpha}{\Omega} = \frac{\Delta\omega}{2\Omega} \quad , \qquad \Delta\omega = \omega_2 - \omega_1 \tag{5.7}$$

Here, $\Delta\omega$ is the 'halfpower-point bandwidth', i.e.

$$A(\omega_1) = A(\omega_2) = \frac{A_{max}}{\sqrt{2}} = \frac{A(\Omega)}{\sqrt{2}} \quad ,$$

or

$$|H^2(\omega_1)| = |H^2(\omega_2)| = ½|H^2|_{max} = \frac{|H^2(\Omega)|}{2}$$

In broadband random excitation, which is not measurable, $\Delta\omega$ is estimated directly as the response spectrum bandwidth at half the maximum value of the spectral density. At this 'halfpower-point bandwidth', the contribution of the damping to the system's dynamic stiffness is the same as that of the detuning from resonance.

This procedure, which does not use any information on phase, is especially convenient when the signal analysis is made solely in analogue form, without appealing to a digital computer, since measurements of phase in analogue form can be much more difficult than those of the amplitude. On the other hand, neglecting information on phase cannot but lead to some loss in accuracy. To improve the estimates, one may use a vector diagram, or Nyquist plot (see Fig. 5.1(b)), which is significantly 'stretched' in the near-resonant zone. Whereas the resonant frequency may be estimated from a Nyquist plot, from its lowest point, the frequencies ω_1 and ω_2, which define the halfpower-point bandwidth, may be estimated with better accuracy from the extreme right-hand and left-hand points of the plot, respectively, so that eqns. (5.7) may be used once again. It may be advantageous also to estimate these points by applying a curve-fitting procedure to the whole Nyquist plot, or a modal circle (since, for a lightly damped system, the shape of this plot is approximately circular), so as to avoid locating them exactly in the course of an experiment.

This approach becomes especially advantageous for MDOF systems with close natural frequencies. In Fig. 5.3, a Nyquist plot is shown for a TDOF

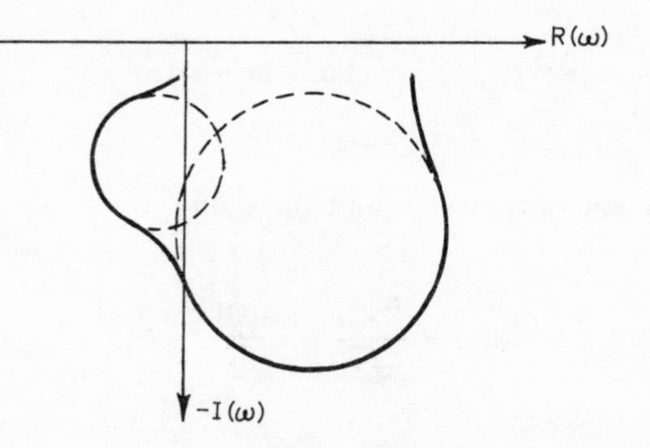

Fig. 5.3 Nyquist plot for a TDOF system with close natural frequencies (solid line) and separate modal circles for these DOFs (dashed lines)

system with close natural frequencies (solid line), together with separate modal circles for the two DOFs (dashed lines). This plot clearly shows that two resonances are indeed present; whereas, in the AFR, only a single resonant peak may be observed, with the second DOF manifesting itself only as a slight 'bump' on the AFR curve. Thus, the amplitude-phase approach provides a better resolution of close natural frequencies, both in a qualitative and in a quantitative sense.

Specific procedures for transfer-function curve-fitting in the frequency domain, both for SDOF and for MDOF cases, are presented in detail in [22]. One of these procedures is particularly appealing to an engineer since it provides a clear representation of the accuracy of fit by transforming a modal circle into a straight line, through the use of the dynamic stiffness rather than the compliance. Various other aspects of modal testing, and of processing and presenting the results of such tests, may be found in [22,52].

Chapter 6
Certain Identification Problems for Vibration Transmission Paths and Excitation Sources

In this chapter, once again linear time-variant dynamic systems, which can be described adequately by their impulse response or transfer functions, are considered. Whenever any pair of participants in the basic trio of excitation-system-response is known, the third one may be found, in principle, by appropriate calculations. The case where the unknown response is sought for is a common one at the design stage of a machine or structure, where the system's impulse response or transfer functions and excitation forces are (presumably) known in most cases from a dynamic analysis and from special measurements, respectively; previous experience may also be used for estimating the excitation and/or damping forces, which are usually the most uncertain characteristics at the design stage. These calculations involve convolution operation(s) in the time domain or simple multiplication(s) in the frequency domain. This design problem is usually called the 'direct' one.

In identification or diagnostic problems, which are sometimes regarded as 'inverse' ones, the response signal is assumed to be known; in fact, it is a basic measured signal. The problem may involve a reconstruction of the input or the excitation signal, with the system's dynamic characteristics being known from other, independent measurements; or, vice versa, the input force may be known, so that the inverse problem involves an estimation of the system's properties. Moreover, both the excitation and the system may sometimes be partly known (some of the input signals and some of the system's transfer functions may be measured in separate tests) and their unknown parts may be sought in this 'mixed' inverse problem.

The procedures for solving such inverse problems may involve a 'deconvolution' in the time domain (a solution of the integral eqns. (5.1) for the unknown $G(t)$ or $f(t)$) or simple algebraic operations in the frequency domain, where, according to eqn. (5.2),

$$F(\omega) = \frac{X(\omega)}{H(\omega)} \qquad \text{or} \qquad H(\omega) = \frac{X(\omega)}{F(\omega)} \tag{6.1}$$

(of course, for multiple inputs and/or multiple measured response signals, there will be a set of such equations). However, the solution to inverse problems faces a basic difficulty: the result may be extremely sensitive to uncertainties in the given data (say, to response measurement noises). The origin of this sensitivity lies in the following general fact: if widely different causes lead to rather similar consequences, it may be difficult to establish exactly the true cause of a certain given result. In mathematical terms, it is said that, if a direct problem is well-posed, then the inverse problem is ill-posed. For dynamic systems, this inequality of the direct and the inverse problems is due to the filtering properties of stable causal systems. For example, the response spectrum of a lightly damped SDOF system to broadband random excitation is dominated, as we have seen in Chapter 3, by the near-resonant components. Away from resonance, the response level is small, and, whenever it is of the same order as the measurement noise, the recovery of the details of the input excitation (at nonresonant frequencies) from the response spectrum data (and presumably a known AFR) can be achieved only with a rather large error.

These basic difficulties, which have been illustrated above for a system with a single input and a single output, naturally become significantly aggravated for multiple input and output (response) signals. The point is that the response signal(s) may not be sufficiently sensitive to certain input excitations, so that the latter are unlikely to be recovered from the response signal(s). Similarly, if certain transmission paths of vibrational energy are mostly circumvented in given tests (or inservice conditions), it may become difficult to identify them from these tests. In all such cases, a set of algebraic equations, representing a multidimensional counterpart to the basic relation (6.1), in the frequency domain, is 'ill-conditioned', with its determinant close to zero; this leads once again to a high sensitivity of the solution of the inverse problem to measurement noise and/or errors.

Similar difficulties may also arise for systems with distributed parameters. Consider, for example, the bending vibrations of a beam, excited by broadband random (timewise) loading with an unknown spatial distribution. Let these measured vibrations be predominantly of a single-mode resonant type. Let the problem be as follows: to use these data for predicting the vibration level of the beam after introducing a certain design modification, such as a variation in the support conditions or an attachment of an additional mass, which leads to a change in the beam's mode shape.

To make such a prediction, one should, in general, estimate the spatial distribution of the load from the available response data for the original design (assuming that the design modification does not change this distribution). The modal response, however, is dependent only on an integral of this distribution along the beam's length, with the modal shape as a weighting function in the integrand. Therefore, an accurate recovery of this distribution from the modal response data is once again quite difficult. To obtain the desired prediction of the response level for the modified beam, one should use some reasonable *ad hoc* assumption concerning the form of the unknown load distribution. (By the way, one approach to this problem is to obtain a conservative prediction by using a 'worst case', rather than the true, loading; this leads to the maximal response level of the modified beam. However, this is impracticable, since such a 'worst case' corresponds to a concentrated load of infinite magnitude, applied at a nodal point of the original beam's mode; if this point is not also a node of the modified beam's mode, the predicted rms response level is also infinite, which is indeed too conservative.)

At present, various general methods and procedures do exist for solving such 'ill-posed' problems [48]. Most of them include some 'smoothing' of the available data; i.e. exclusion of components with too high frequencies. For the recovery of multiple input excitations, the procedures for 'multichannel inverse filtering', described in [42], may be useful. An example of a successful recovery, for a certain case of a single input and a single output, is presented in [34]. The problem was to obtain the combustion pressure pulse in the cylinder of a Diesel engine from a vibrational signal, as registered by an accelerometer attached to the cylinder casing. This acceleration signal had a fairly broad spectrum, owing to the strong magnification of the high-frequency components by the numerous resonances of the casing; the latter could easily be seen in the frequency-response curves, $A(\omega)$ and $\varphi(\omega)$, of the transfer function between the dynamic force within the cylinder and the casing acceleration. In this particular case, the recovery of the original input combustion pressure signal was based on a direct 'inverse filtering' of the measured vibration signal by a filter with frequency responses $A^{-1}(\omega)$ and $-\varphi(\omega)$. A decisive success for this inverse filtering procedure is claimed [34]: the pressure waveform was recovered with such an accuracy that it became possible to detect changes in the combustion process due to a leak in the injector pump.

It seems that this success in monitoring the pressure oscillations within a cylindrical shell, by measuring the shell's vibration, may also be important in other industries. However, this direct inverse filtering technique presents stringent requirements concerning the accuracy of the available frequency-response data. This is similar, roughly speaking, to the successive multiplication and division of a certain quantity by the same big number, which may not lead to the original result because of round-off errors. In the same way, successive passage of the input signal first through a 'direct' filter (i.e. a given system, or cylinder casing in the above example) and then through an inverse one can make the result too sensitive to errors and noises, if each of these filters introduces significant variations in the signal. Such variations are indeed present, owing to the many casing resonances, and they may lead to distortions in the recovered signal; in particular, strong spurious high-frequency components may be obtained. Therefore, some smoothing procedure may be advantageous that makes the recovered signal less sensitive to variations in given data. One such procedure, cepstral analysis, briefly described in the following, has been shown [34] to provide this desired property of the recovered signal, for the above example of a Diesel-engine cylinder.

Cepstral analysis is a specific nonlinear filtering procedure, which may be used with advantage for solving both of the basic identification problems considered in this chapter; namely those of identifying either the vibration transmission paths or the excitation sources. Its advantageous properties are more of a qualitative than a quantitative nature, and, in general, it may often make the desired solution to a given inverse problem less sensitive to uncertainties in given data.

The basic operations of cepstral analysis are as follows: a Fourier transform (FT), taking the logarithm of the latter and then an inverse FT. When applied directly to a signal, x(t), these operations yield the complex cepstrum $C_X(\tau)$ as an inverse FT of $\ln|X(\omega)| + i\arg X(\omega)$, where $X(\omega)$ is the FT of the signal x(t), with amplitude and phase $|X(\omega)|$ and $\arg X(\omega)$, respectively. The above operations may also be applied to the correlation function $K_{XX}(\tau)$ of a stationary random process x(t). In this case, a power (real) cepstrum $\hat{C}_{XX}(\tau)$ is obtained as the inverse FT of $\ln \Phi_{XX}(\omega)$, where $\Phi_{XX}(\omega)$ is the spectral density of x(t). The title 'cepstrum' emerged, in fact, as a result of transporting several letters in the word 'spectrum'.

Now, what is the point of returning to time domain, after the nonlinear logarithmic transformation? First of all, the following property of the logarithmic transform is important: since the derivative of a log function $\ln|X|$ is inversely proportional to $|X|$, this transformation should cut short the 'tails' of $|X(\omega)|$, when this curve has the shape of an AFR of an SDOF system. However, the most important feature of cepstral transformation is, perhaps, the reduction of a measured response signal to such a form that it becomes a simple sum of the additive contributions of the input excitation and the system's dynamic characteristics. This can easily be seen from eqn. (5.2), since

$$\ln X = \ln H + \ln F \tag{6.2}$$

Therefore, applying an inverse FT to eqn. (6.2) yields:

$$C_X(\tau) = C_h(\tau) + C_f(\tau) \tag{6.3}$$

where $C(\tau)$ are the cepstra of the functions indicated by the corresponding subscripts. If any pair of signals has its cepstra localised in disjoint regions in the time domain, these two signals may be separated easily by simple temporal windowing, i.e. selecting only parts of the composite signal at certain finite time intervals; no detailed information on these component signals is needed, as long as they do not overlap. This may be regarded as a time-domain counterpart of bandpass filtering, which is efficient for additive signals with nonoverlapping frequency spectra.

This property of cepstral analysis may be quite helpful, therefore, in the 'deconvolution' of the system and the input signal, since it reduces the overlap of the component signals. Various examples of this kind are presented in [34]. Some of these, formulated in acoustical terms, involve isolating the directly propagated or primary signal from the echoes; i.e. the various multiple reflected signals. Since acoustical signals propagate without dispersion, and thus preserve their form (see Chapter 4), cepstral analysis may provide a complete straightforward solution to the identification problem for this specific case: the separated components of the measured response signal directly represent the primary and the reflected signals. This is not so with the previously considered example of a Diesel-engine cylinder casing. Here, the medium is dispersive, because of the predominantly bending vibrations of the casing, so that cepstral analysis by itself may be insufficient for a complete recovery of the input signal, without inverse filtering; in view of the possible modal representation of the cylinder casing, this example may be regarded as a rather general one of a dynamic system with lumped parameters. However, cepstral analysis indeed proved helpful in smoothing the available data, so that the inverse filtering became 'robust'; i.e. the results became rather insensitive to the specific choice of inverse filter [34]. Examples of applications of cepstral analysis also include cases where the excitation signals are in the form of periodic pulses; once again, these may be more readily separated from the system's response and/or from other periodic signals, with different periods [34].

Consider now two general identification problems, which are quite typical for complex structural systems and machines. They are illustrated by the flow charts in Fig. 6.1 (three 'channels' are shown, for each case, but it is obvious

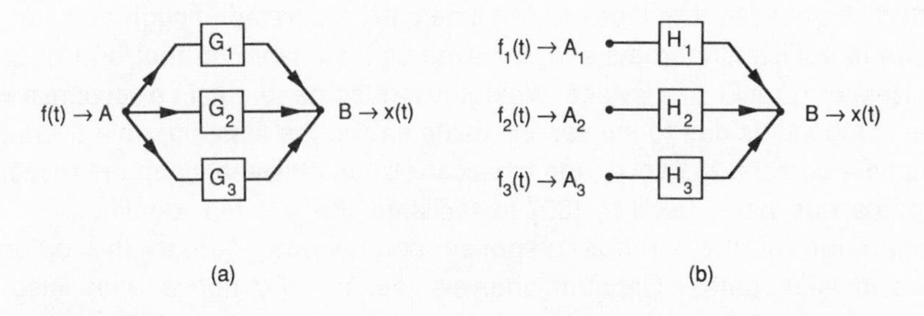

Fig. 6.1 Flow charts of two identification problems: evaluation of vibration transmission paths (a) and separation or localisation of excitation sources (b)

that, in principle, any finite number of such 'channels' is possible).

(a) Evaluation of the vibration transmission paths (Fig. 6.1(a)). A system with a single input A and a single output B is considered. This system may be described, of course, by its overall transfer function. Assume, however, that a somewhat deeper insight into the system's behaviour is desired. Thus, the excitation signal may propagate to the output B through various paths or 'channels', with different impulse response functions $G_i(\tau)$. The problem is to estimate, from measurements of both the input and the response signals, how the total response energy is distributed between these different paths. The possibilities for practical application of such estimates are obvious: they may show which of the transmission paths are mostly responsible for undesirable noise and/or vibration at point B and thus should be the first ones to be dealt with whenever vibration control is needed.

(b) Separation or localisation of the excitation sources (Fig. 6.1(b)). The response signal, x(t), at a single output B is made up of the contributions due to several excitation sources. The latter produce signals $f_i(t)$ which arrive at B after passing through separate paths or channels with transfer functions $H_i(\omega)$. The sources are assumed to be noninteractive and additive; i.e. their contributions are independent of one another and are simply summed up at B. The problem is to estimate, from measurements of $f_i(t)$ and x(t), the contribution of each source to the overall measured response signal. The practical applications are obvious, once again, as for problem (a): such estimates may show which source(s) should be suppressed in the first place to provide the most effective noise and/or vibration control.

The first of the above problems has been considered in detail in [30] for tests with a controlled input excitation. Separation in the time domain was used, based on differences between the transit times for signal propagation from input to output along different paths. This approach may indeed be effective provided that these transit times are separated enough and that the input is sufficiently broadband; for example, excitation in the form of short pulses, or a rapid sine sweep, may be used, so as to obtain a response with the components due to the various paths having the least possible overlap in the time domain. Moreover, the crosscorrelation between input and response signals has been used in [30] to facilitate the desired identification and separation of the various response components due to the different transmission paths. Cepstral analysis, as outlined above, may also be advantageous in this connection. Indeed, it should lead to a 'contraction' of the various components of the crosscorrelation functions, which correspond to signal transmission through the different paths, and to a reduction of the overlap between these components in the (transformed) time domain after the cepstral transformations.

The systems considered in [30] consisted of resonant SDOF elements, simulating machine components, and pure delay elements, simulating sound propagation in air (see Chapter 4); the case of beam elements with dispersive bending waves was also considered. The identification and separation of the various paths' components for nondispersive cases were based on the maxima of the crosscorrelation functions, as in the procedures for the incident, reflected and transmitted signals in one-dimensional systems with distributed parameters, described in Chapter 4. If the system contains resonant element(s), however, the signal should be broadband compared to the AFR of the latter; otherwise, as shown earlier in this chapter, the crosscorrelation function may not be sufficiently asymmetric for the dynamic characteristics of the system to manifest themselves.

Paper [30] contains some useful formulae for a rapid sine sweep. For dispersion due to beam elements with bending, or flexural, waves, another, more complicated, type of excitation is preferable to the rapid sine sweep. These procedures can be extended to on-line measurements with 'natural' random input signals, rather than specially produced ones, provided that they are sufficiently broadband. Cepstral analysis may also be used to reduce overlap between the components of the crosscorrelation function, corresponding to the different transmission paths.

Turning to problem (b), consider first a special case, which was analysed as far back as in the mid-fifties [26]. The problem is that of noise propagation from several sources in an acoustic medium; it is assumed that only primary signals may be considered, whereas reflected waves may be neglected, so that each ith 'channel' between the noise source at A_i and the noise detector at B is, in fact, a system with a pure delay θ_i. The sources are assumed to be not only additive but also independent, in the sense that a transducer at point A_i senses only the signal $f_i(t)$ from the ith source whereas its response to other noise sources may be neglected.

For this simple case, the solution is straightforward if a crosscorrelation analysis is applied. Indeed, if the attenuation of the sound waves between the sources and noise detector may be neglected, the response signal x(t) is simply a sum of the signals $f_i(t-\theta_i)$, reproducing the excitation signals f_i with time delays θ_i. Then, since all the f_i are uncorrelated, the absolute maximum of a normalised crosscorrelation function:

$$r_i(\tau) = \frac{\langle f_i(t)\, x(t+\tau)\rangle}{(\langle x^2(t)\rangle \langle f_i^2(t)\rangle)^{1/2}} = \frac{\langle f_i(t)\, f_i(t+\tau-\theta_i)\rangle}{\left[(\sum_{j=1}^{N} \langle f_j^2(t)\rangle)\, \langle f_i^2(t)\rangle\right]^{1/2}} \tag{6.4}$$

provides the desired estimate of the ith source contribution to the overall rms noise level $\langle x^2(t)\rangle$. Indeed, at $\tau = \theta_i$, the numerator in eqn. (6.4) is simply the mean square of the ith noise signal $f_i(t)$, so that:

$$r_i(\theta_i) = \left[\frac{\langle f_i^2(t)\rangle}{\sum_{j=1}^{N} \langle f_j^2(t)\rangle}\right]^{1/2} = r_{i,max} \tag{6.5}$$

Since the mid-fifties, many researchers have considered this problem under less restrictive assumptions. First of all, the assumption of independent sources may be relaxed, so that $f_i(t)$ are assumed to be correlated. The separation of the sources is fairly easy in this case provided that the transit times θ_i between the excitation sources and the response detector at B are sufficiently distinct and/or the noises are sufficiently broadband, so that their autocorrelation functions decay rapidly and thus do not overlap after they are combined in the crosscorrelation functions $\langle f_i(t)\, x(t+\tau)\rangle$. In general, however, the system may not have well defined transit times for the propagation of signals from the various inputs to the point where the output or response signal is measured. The problem may, of course, be solved by simple bandpass filtering provided that the spectra of the various excitation sources lie within nonoverlapping frequency bands. It is probably this sort of algorithm that is used by a worried car owner who is listening attentively in an attempt to identify possible sources of an unfamiliar noise within his car.

The general case may be analysed in the frequency domain provided that all the transfer function $H_i(\omega)$ between the inputs A_i and the output B are known from independent tests or theoretical predictions. Well established procedures for such analyses do exist [4,5], which use multiple and partial coherence functions.

We conclude this chapter with a practical application in nuclear power engineering, for which a simple crosscorrelational analysis of the responses was successful, even in the absence of independent detailed information on the system's dynamic characteristics. In Chapter 7, another extension of problem (b) is considered; namely to a case of interactive or nonadditive excitation sources.

Figure 6.2 is a schematic of a network of pipes, feeding water to a Reactor of High Power of Channel type (RBMK)*. This item, the first one, underwent 'cold' hydraulic tests at the Leningrad Nuclear Power Plant in the early seventies. The tree-like network consists of a main head pipeline (MHP) with a built-in main circulation pump and distributing group collectors (DGC) of smaller diameter attached to it; to each of these group collectors, 46 lower water pipelines (LWP) — fairly flexible tubes of yet smaller diameter (57 mm) with a 90° bend — are attached. Thus, altogether about 1000 LWPs are used to feed the cooling water separately into the active zone of the reactor. The scheme shown in Fig. 6.2 is referred to both in this chapter and in subsequent ones.

During hydraulic tests, vibrations of the pipes were measured. These vibrations were extremely low, and did not affect the plant reliability. Nevertheless, it was decided to complete the experimental programme as planned, since the vibration signals had been recorded; however, pressure oscillations within the pipes could not be measured in these tests.

Figure 6.3(a) shows the spectral density of the LWP vibrational signal. It contains three quite sharp peaks, with the intermediate one at a frequency 2.8 Hz corresponding to the resonance of the lowest out-of-plane mode of the LWP. Figure 6.3(b) illustrates the crosscorrelation function of this signal and that of a strain gauge attached to the wall of the main head pipeline. Two narrowband components with mean frequencies of about 0.25 and 4.6 Hz can be seen. Evidently, they correspond to the extreme left-hand and right-

* This abbreviation corresponds to the Russian title.

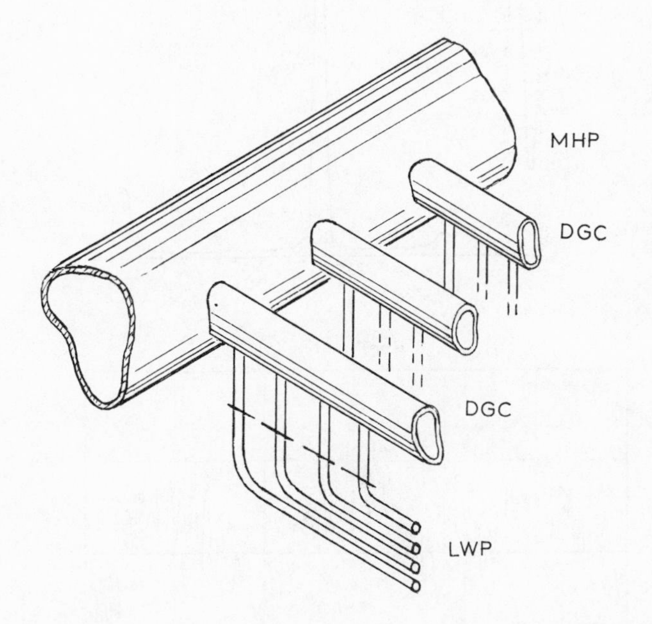

Fig. 6.2 Sketch of the low water pipeline (LWP) network of the RBMK (Reactor of High Power, Channel-type). The horizontal dashed line is a rigid tie for vibration control

hand peaks of the spectrum in Fig. 6.3(a), whereas the resonant component at 2.8 Hz is not present. Thus, the latter component of the LWP vibration is excited by some (presumably hydrodynamic) forces within the tube itself, whereas the two other frequency components are excited by disturbances propagating from the main head pipeline. So, the contribution of the 'internal' forces within LWPs to their overall vibration level is the ratio of the area under the middle peak of the spectral curve in Fig. 6.3(a) to the area under the whole curve. The rest of the LWP vibrations are excited by the motions of the DGC.

The aims of this analysis were related to the prospects of vibration control. At the design stage of the RBMK, the engineers were concerned with the

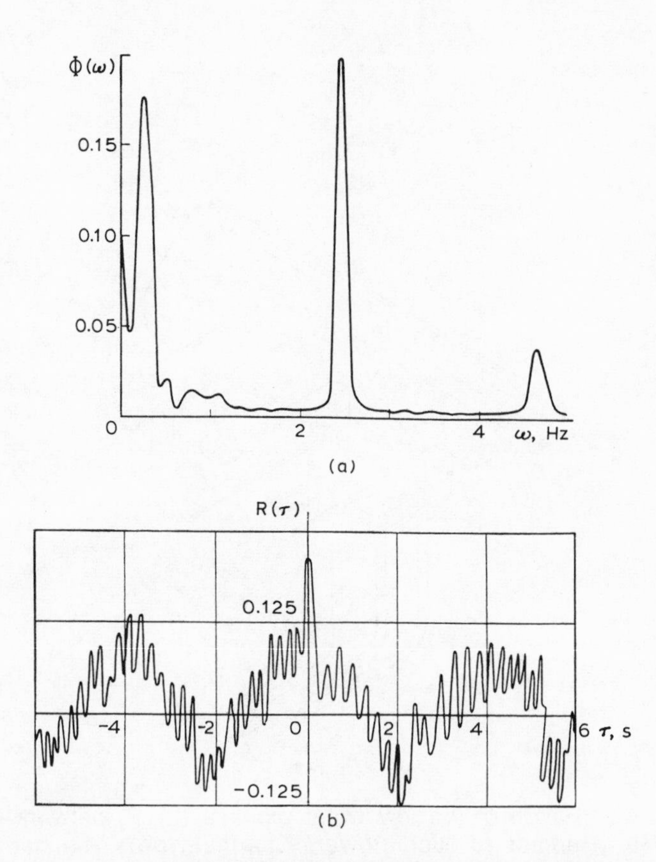

Fig. 6.3 Vibration of RBMK pipes: spectral density of the LWP response (upper graph) and crosscorrelation function of the vibrational signals from the MHP and LWP with the two narrowband components corresponding to the first and third peaks of the spectral density (lower graph)

possibility of strong flow-induced vibrations of the LWPs, which had a high flexibility in the out-of-plane direction. So, it was decided to provide measures for vibration control, whenever required. Since external additional stiffening components cannot be attached to LWPs, without interfering with the complex pattern of their thermal expansions, it was decided to apply rigid 'ties' to join all the LWPs, attached to the same DGC, into a single bundle. The basic idea was to obtain a more uniform distribution of vibrational energy between the various LWPs, rigidly connected one with another.

This approach to vibration control may be quite efficient for a sufficiently high number of elements within a bundle, provided that each element of the set (i.e. each LWP) is excited independently by its own internal force. On the other hand, this approach may be rather inefficient for correlated or synchronous excitation forces; in the present example of LWP vibrations, this case corresponds to synchronous 'kinematic' excitation due to vibrations of the DGC or the 'common foundation' of the LWPs. (Variations in the natural frequencies among the LWP set due to differences in their dimensions may also contribute to improving the uniformity of the energy distribution within the bundle of LWPs; however, for simplicity, this effect is not considered here.)

Simple quantitative analysis shows that, for uncorrelated excitations due to 'internal' forces in the LWPs, the application of a rigid tie should lead to an approximately (N/4)-fold reduction in the maximal mean-square vibration level, where N = 46 is the number of similar elements (LWPs) combined into a bundle. However, for 'synchronous' excitation from the DGC, the corresponding reduction factor is not much higher than unity.

Thus, the solution to the problem of separating the excitation sources may be used to predict the efficiency of the vibration control of the set of similar components (LWPs) by rigid ties; i.e. to estimate the vibration reduction factor æ. Its overall value is a weighted sum of the 'internal' and 'external' relative contributions to the total mean-square response, the weighting factors being N/4 = 11.5 and (almost) unity, respectively. These conclusions were verified in the above hydraulic tests, some of which were made with the rigid ties installed. As already mentioned, these studies were 'just for fun', since the measured vibration levels of the RBMK pipes were quite small. It is possible, however, that the results of such an analysis may be applied to some other systems, for which the development of similar vibration-control methods may be crucial.

Chapter 7
Estimation of the Stability or Reliability Margin and Identification of Parameter Variations from On-Line Response Data

This chapter deals with the semiqualitative identification of dynamic systems [21]. The two preceding chapters were mostly concerned with the complete quantitative identification of linear, time-invariant systems; i.e. with the determination of all their dynamic characteristics, and an estimation of all the parameters of the system, and/or the excitations etc. However, as the last example of Chapter 6 shows, such a complete identification may sometimes be both unnecessary and impracticable. In fact, for some applications only a few key parameters may be required; in the above example of RBMK pipelines, it was, in fact, a single parameter — namely the ratio of the mean-square responses of the LWPs to the 'internal' and 'external' excitations. On the other hand, the basic input information required in general for separating the excitation sources, or for solving the decomposition problem (such as the complete set of transfer functions), may not be available for a given specific case; indeed, it was quite difficult to measure the transfer functions of the RBMK piping in direct infield testing. In some cases, an accurate complete identification may be difficult because the system is time-variant and/or nonlinear.

Thus, a semiqualitative approach to identification, as well as a purely qualitative one (see Chapter 8), may emerge naturally as an inevitable tradeoff between the availability of input data (which may be limited in proof tests, particularly for inservice measurements in complex machines or structural systems) and the desired information on the internal state of a given system.

The first problem to be considered is estimating the stability margin (or the reliability margin, in general) of the system from its on-line response data. It is obviously important for people, living near a sleeping volcano, to know whether an eruption is imminent; or for a blind person to know how close to the brink of a precipice he is walking. In a machine or structure, instability may lead to the collapse of a bridge truss or a shell-type aircraft structure, or to excessive detrimental vibration in an aircraft wing, a rotating shaft of an axial compressor, or a tube array of a heat exchanger. In such cases, it is useful, and sometimes essential, to provide more or less continuous inservice monitoring of the system's stability margin; i.e. to estimate how close the machine or structure is to an instability threshold.

The simplest example of static instability is a slender and flexible steel strip, loaded by compressive forces at its ends. When the compression force is sufficiently high, the strip 'buckles'; i.e. bends laterally, even in the absence of any apparent lateral loading. The reason is that, above the critical compressive force, the strip becomes unstable*, i.e. its flexural resistance is completely exhausted. Then, even vanishingly small disturbances may lead to a significant bending of the strip, which then progressively increases because of the high compressive force. Such an instability of an important compressed slender element, say of a bridge truss, may indeed lead to a catastrophe, a real one rather than many of those studied in Catastrophe Theory in mathematics.

It is obvious that, since a compressed slender strip or beam buckles because its bending stiffness becomes exhausted completely owing to a compressive loading, the value of the fundamental bending natural frequency of the beam may serve as a good index of the stability margin. The higher the compressive load, the lower this natural frequency, and at the critical load, or stability threshold, it is reduced to zero. Therefore, inservice monitoring of the static stability margin is possible, in principle, for such components, based on a measurement of this natural frequency. This can be done continuously whenever sustained inservice excitation is present, causing small bending oscillations in the component. While this response is small, and presents no danger, it may, when treated properly, be used as a 'friend': whenever its lowest spectral peak shifts too close to the origin, it gives a warning of a dangerous approach to the critical load. (In general, this should refer to the peak which is clearly identified as corresponding to the fundamental natural frequency.) A possible conceptual example is an element of the skin of the upper (compressed owing to lift forces) surface of an aircraft wing, the necessary external inflight excitation being provided by pressure oscillations in the turbulent boundary layer and/or by atmospheric gusts.

Problems of static stability are of great importance in some structural systems, such as stiffened shells, used in aircraft or submarines. These problems, however, are fairly well understood, and many sophisticated methods and computer programs exist for static stability analyses of structures

* To be more rigorous, the term 'unstable' should be applied to the initial straight shape of the strip.

at the design stage. However, inservice stability-margin monitoring is much more important with respect to another type of instability — the dynamic one.

In Chapter 3, several examples were presented of dynamic instability phenomena, namely of a sudden drastic growth of the vibrations in a machine or structure. Figure 7.1 provides a qualitative description of a

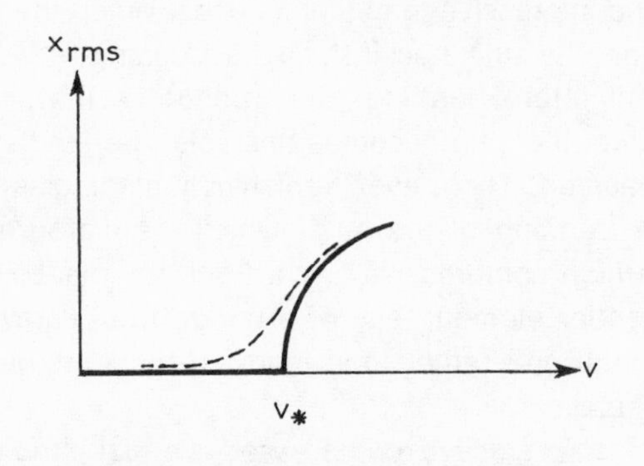

Fig. 7.1 RMS dynamic displacement as a function of a certain operational parameter v in a system without any external excitation (full line) and in the presence of some external excitation in the subcritical state (dashed line); v_* is a stability threshold

system's behaviour when a certain basic operational parameter v is varied. The latter may represent the flight velocity of an aircraft or the flow velocity in wind-tunnel flutter tests of its model; the crossflow velocity in a tube array of a heat exchanger; the rotation speed of a circular saw; the travelling speed of a railway car; etc. The ordinate in Fig. 7.1 represents a typical dynamic displacement of the system, such as its rms value, x_{rms}, at a certain point. The solid curve corresponds to the case where no sustained external excitation is present, and therefore no response can be observed in the system in the subcritical state, i.e. as long as the value of the basic parameter v is less than a certain threshold or critical value v_*. However, whenever the latter is exceeded, the system starts to oscillate violently.

Of course, in some applications, more operational parameters may be relevant, so that for, say, two such parameters, the instability boundary will be represented by certain line(s) in the plane of these parameters, rather than by a single threshold value. For example, the dynamic stability of a metal-cutting process may depend on two basic operational parameters, namely the depth of the cut and the feed rate; and of course the dynamic properties of the lathe may also be important. Certain combinations of values of the above parameters may correspond to dynamic instability and self-excited oscillations in the whole system lathe-tool-workpiece. This phenomenon, known as regenerative chatter, may lead to a deterioration in the cutting process, with a wavy shape of the machined surface.

Systems may behave differently after the stability threshold is exceeded. In many cases, the level of the system's oscillations in a supercritical state ($v > v_*$) is limited by various nonlinearities. Such systems may operate in the regime of these self-sustained oscillations. This 'mild' instability may be undesirable, of course. For example, the oscillations of a railway car due to its lateral instability may have a limited amplitude because of the strong nonlinear interaction of the wheels and their rim flanges with the rails; however, these oscillations may be uncomfortable for the passengers. As another example, the tubes of a heat exchanger, after becoming unstable in a crossflow, may oscillate with a limited amplitude owing to the restricting effect of their intermediate supports with small gaps, or tube plates. These vibrations lead to a reduced life, because of fretting wear, but not to an immediate failure. Since, for such systems, operation in a supercritical regime ($v > v_*$) is not completely prohibitive, various diagnostic problems may be of interest which involve both the sub- ($v < v_*$) and the supercritical ($v > v_*$) regimes. Some of these problems, such as the discrimination between the two different regimes from on-line response measurements, are considered in Chapter 8.

On the other hand, structural nonlinearities in many systems cannot restrict the growth of vibration amplitudes in the supercritical state to a more or less safe level. This implies that such oscillations may lead to an immediate, complete failure or to the breakdown of the system, soon after the stability threshold is exceeded. Flutter of an aircraft wing and/or tail unit is a well known example of such a catastrophic instability, which was lethal for many test pilots of early jet planes. This instability-related vibration is a foe indeed, and a very dangerous one since it may strike without any warning: according to the full line in Fig. 7.1, everything is calm when $v < v_*$; however, even incrementally exceeding the threshold level v_* (critical flutter speed in the above example) may have disastrous consequences.

Obviously, a diagnostic system providing an on-line warning of a dangerously close approach to the instability threshold would be most useful. Of particular interest is the case where certain measurable oscillations are observed in the subcritical regime of the system's operation, as illustrated by the dashed line in Fig. 7.1. These inservice oscillations, due to some 'natural' excitation, may be small and within safe limits (at least, as long as v_* is not approached too closely). Moreover, when processed properly they also may be used as a friend, warning of the approach of the dangerous foe, by directly estimating the stability margin from these data.

Of course, the current state-of-the-art of the design of potentially unstable systems, without diagnostic facilities, is not so gloomy as indicated in the above paragraph. Stability thresholds are usually predicted in terms of the basic operational parameter(s), at the design stage, and these predictions are then verified in proof tests. However, such a threshold may change subsequently, owing to some inservice variations in the system's properties; moreover, variations in this threshold within a set of nominally identical units may also be possible because of inevitable manufacturing tolerances. Whenever all such deviations from the baseline stability limits are of concern, diagnostic facilities, which provide an inservice monitoring of the stability margin in terms of the system's dynamic characteristics (rather than in terms of the measurable operational parameters), may be of high importance; the stability margin estimate will then be valid for the specific unit rather than for a 'nominal' one, as introduced according to design.

Moreover, such diagnostic facilities may also be used in those basic proof tests that are used to check and finally establish the stability boundaries in terms of the system's operational parameters. Thus, in the wind-tunnel flutter testing of an aircraft, or its scale model, there is the problem of saving the expensive model from fracture, which may become possible whenever the critical speed v_* is exceeded. A possible solution is to somehow excite oscillations of the model in the subcritical state and then to estimate v_* by the appropriate processing of the measured subcritical response. Of course, whenever such an excitation already exists owing to, say, the natural turbulence of the flow in the wind tunnel, it may be used with advantage to directly estimate the stability margin in terms of the system's dynamic characteristics, as identified from its subcritical response. A series of such measurements for a fighter plane [43], at various flow speeds, has been used to estimate the critical flutter speed by an appropriate extrapolation of the relation between the flow speed v (at $v < v_*$) and the stability margin, as estimated from subcritical response measurements.

The stability margin, as obtained in terms of the system's dynamic characteristics, may sometimes be used directly, without recourse to measurable operating parameters. First, it may be used for comparative purposes; thus, one of a pair of similar systems, such as helicopter rotor blades, may be regarded as more stable than the other one if the response to external pulse-like disturbances, say, of the flapping type, decays more rapidly, provided that the relation between this decay rate and the basic variable operational parameter is monotonic (at least in the vicinity of the stability threshold). The latter condition is usually satisfied, and may be verified by basic tests and/or field experience, which may also be used to establish safe limits for the system's stability margin, as expressed in terms of its dynamic characteristics. Finally, a stability margin may be used as an index for the system's performance, which is directly related to its sensitivity to various disturbances. Thus, for example, in [37] a clear experimental correlation was established between the stability margin of circular saws in terms of their rotation speed and the accuracy of the cut, which governs the relative volume of wasted material.

The simplest case of dynamic instability is that in which only single-mode oscillations are involved. This case may be referred to as that of a negative apparent modal damping, so that the minimal model damping ratio may be regarded as the system's stability margin; in certain cases, it may also be the damping ratio of a certain specific mode that is known from field experience to be the most dangerous one. This problem of estimating modal damping ratios for a linear time-invariant system has already been studied in Chapter 5. In the following, we present more details for a broadband random excitation, which cannot be measured directly.

For an SDOF system, or a well separated spectral peak in an MDOF system response, the damping ratio may be estimated either from the spectral density of the response (for example, from its halfpower bandwidth) or from the decay rate of the envelope of the response autocorrelation function. This autocorrelation function, $K_{xx}(\tau)$, is estimated from a response sample, either directly, as described in Chapter 2, or using the 'Randomdec' algorithm [50]:

$$D_{x0}(\tau) = \langle x(t+\tau) \mid x(t) = x_0 \rangle = \left[\frac{K_{xx}(\tau)}{K_{xx}(0)} \right] x_0 \tag{7.1}$$

Thus, $D_{x0}(\tau)$ is easily estimated from a given response sample as a conditional expectation of $x(t + \tau)$, the condition being $x(t) = x_0$, where x_0 is a certain chosen threshold level. From eqn. (7.1), the modal damping ratio is estimated directly from the decay rate of $D_{x0}(\tau)$. Both these methods (Randomdec and halfpower bandwidth) were used to estimate the stability margin of an aircraft wing, from torsional response measurements, in the above wind-tunnel tests [43].

For an MDOF system, a preliminary bandpass filtering may be required to extract the 'useful' modal signal from the overall measured multimodal response. We consider only the case of well separated spectral peaks. It is then possible to make the filter bandwidth small compared with the peak spacing, so as to achieve a sufficient filtering-out of other modal responses, and high compared with the modal bandwidth, so as to produce least distortions of the 'useful' modal response signal. Such distortions are, however, inevitable to some extent, and it is extremely important for the estimation procedure to be robust, or 'self-correcting', with respect to these distortions.

One possible approach, which is also very clear, is to use a semilogarithmic plot of the above envelope $\bar{K}_{XX}(\tau)$ of $K_{XX}(\tau)$. For an SDOF system, this should yield a straight line with a slope $-\alpha$, since

$$-\ln \bar{K}_{XX}(\tau) = \alpha\tau \qquad (7.2)$$

The value of α may be estimated, not from the whole plot of $-\ln \bar{K}_{XX}(\tau)$ but rather from that part that is actually linear. Thus, the influence of bandpass filtering, which usually leads to deviations from a straight line at small time shifts τ, can be avoided; the region of high random errors of the K_{XX} estimate, at high τ, may also be excluded from the curve-fitting procedure. Moreover, deviations of $-\ln \bar{K}_{XX}(\tau)$ from a straight line may also be used to detect a second spectral peak, left unnoticed in the course of a preliminary rough spectral analysis.

This procedure works quite well as long as the system is indeed linear. However, it definitely is not robust with respect to another inadequacy of the basic linear SDOF model, namely to a small nonlinearity of the system's restoring force. The latter may lead to a significant broadening of the response spectral density of a lightly damped system. Indeed, the response in the nonlinear case may, in fact, be regarded, roughly speaking, as a collection of cycles, with the frequency $\omega(A)$ of each cycle being dependent on its amplitude, A, rather than a constant. This spectrum-broadening is of the order $\omega(\langle A\rangle) - \omega(0)$; this implies that, if the damping ratio is, say, about 1%, the 'natural' spectral peak of the linear SDOF system's response will be overlapped owing to this nonlinear effect whenever the difference between the values of natural frequency at the mean amplitude and at vanishingly small amplitudes is higher than 2 or 3%. Therefore, even such a small nonlinearity of the restoring force may lead to significant errors in the damping estimates as obtained by a direct correlational or spectral analysis. Such a spectrum-broadening has been observed, for example, in the numerical simulation of an SDOF system with a 'flapping' crack (see Chapter 9).

An appropriate procedure, which excludes this nonlinear effect almost completely from the analysis, is based on extracting the amplitude $A(t)$ (or its square $V = A^2$) of the response. Then the effect is essentially 'absorbed' by the response cycles, which are nonisochronous, whereas its influence on the slow fluctuations of the amplitude is of minor importance, for a moderate nonlinearity. The correlation function of the zero-mean part $V_0(t) = V(t) - \langle V\rangle$ of the squared amplitude of the linear response is [21]:

$$K_{v_0 v_0}(\tau) = K_{v_0 v_0}(0) \exp(-2\alpha\tau) \quad , \quad \tau \geq 0 \tag{7.3}$$

so that the damping ratio may be estimated from the plot of $-\ln K_{v_0 v_0}$, once again within its straight-line part. Moreover, the verification of a basic linear time-invariant SDOF model is also obtained if the slope of the $-\ln K_{v_0 v_0}(\tau)$ curve is twice as high as that of $-\ln \overline{K}_{XX}(\tau)$. On the other hand, this basic model in general may not be adequate whenever this relation does not hold — either because of nonlinearity or because the system's natural frequency is in fact time-variant; the latter effect is considered later.

Whenever an MDOF dynamic system becomes dynamically unstable owing to nonconservative loading, two modes at least are involved in the resulting oscillations, namely those with coalesced natural frequencies (see Chapter 3). The term 'binary flutter', from aeronautics, may be used for such numerous, basically TDOF, phenomena. While the critical modal damping ratios may still be used as stability margins, as in [43] for flutter tests on a fighter-plane wing, another type of stability-margin estimate from the subcritical response may be considered. Thus, at the stability boundary (i.e. in the state of neutral dynamic stability), the responses of both the basic modes to any perturbation are periodic with the same period (coalescing of natural frequencies!) and a constant relative phase shift. In other words, the coupling between the two basic modes owing to nonconservative forces reaches here such a level that it leads to completely synchronous motions of the system (at least in the basic DOFs which become dynamically unstable). On the other hand, in the absence of nonconservative loading, these basic modes are usually uncoupled, and therefore their responses to external excitations, which are generally only slightly correlated, are statistically independent. With increasing nonconservative loading, these subcritical modal responses $x_i(t)$, $i = 1, 2$, become more correlated. Finally, at the threshold barrier v_* they should become completely correlated, with a unity crosscorrelation factor $\rho_{12} = \langle x_1 x_2 \rangle / (\langle x_1^2 \rangle \langle x_2^2 \rangle)^{1/2}$. Therefore, the quantity $1 - \rho_{12}$, or $1 - \rho_{12}^2$, may also be regarded as a stability margin, which may be estimated very easily from measured signals $x_i(t)$ by simple, purely analogue, processing. This approach is possible only if at least two modes, or DOFs, participate in a dynamic instability. A similar idea was used in the above-mentioned wing-flutter tests [43], except that the cross-spectral density rather than the crosscorrelation function was measured, with $x_i(t)$ corresponding to the bending and torsional responses of the wing.

The problem of estimating a stability margin from on-line response data may also be considered for time-variant linear systems, or for systems with parametric excitation, where parametric instability (see Chapter 3) is possible.

This problem is studied in the rest of this chapter, together with the problem of estimating the level of the parametric amplification of the response to external excitation. The latter problem may be regarded as a special case of a general excitation-sources-decomposition problem with inherently nonadditive sources (its solution is closely related to that of the problem of estimating the stability margin). The general formulation of this problem (see Chapter 6), which corresponds to the natural concerns of an engineer, has a clear and unambiguous sense only for those cases where the superposition principle holds; i.e. where the responses of a given system to various excitation sources are additive. In many systems, however, the contribution of an excitation source to the overall response may be significantly dependent on the contribution of other source(s). In such a case of 'interactive contributions', a clear definition should be given for the contribution of each source.

In linear systems, with both external and parametric excitation, simple additive superposition of the responses to these two excitations naturally doesn't hold. This can be seen simply from the fact that, in the absence of any external excitation, the steady-state response of a stable linear system is identically zero. In other words, the contribution of the parametric excitation is strongly dependent on the presence of an external excitation. Therefore, the problem is formulated here in the following way: to determine, by an appropriate analysis of the measured overall response, what response level would be observed in the system in the absence of any parametric excitation. This identification problem is that of estimating the parametric amplification of the response to an external excitation. The possible applications for such an identification method are obvious. If excessive vibrations in a given system are observed during its tests or service, the corresponding identification algorithm may provide an estimate of the effectiveness of those design modifications that alleviate the source of the parametric excitation.

These semiqualitative identification problems are considered for stationary broadband external random excitation, which leads to a nonzero subcritical response of a stable linear system. The latter may have time-variant natural frequency(ies), and the cases of periodic and random variations are studied. Both problems should be closely related, since parametric amplification of the response to external excitation should, in general, increase as the parametric instability threshold is approached. In [21], rigorous mathematical derivations can be found of the algorithms, which are described and explained below mainly in physical terms. Moreover, computer simulation studies are described in [21], which were used both to verify the theoretically derived algorithms and to extend them to certain cases that were found to be not amenable to theoretical analysis.

The first case is that of an SDOF system with near-resonant periodic parametric excitation, i.e. the system's natural frequency oscillates periodically with a frequency close to a doubled mean value of the natural frequency itself. As described in Chapter 3, such variations may indeed lead to the system's dynamic instability with the boundaries of the instability domains being presented usually in terms of the amplitude and frequency of parametric excitation (Ince-Strutt chart). Thus, we consider the problem of estimating the stability margin from the system's subcritical response to external random excitation. The periodic parametric excitation affects both the amplitude and the phase of this subcritical response. However, the major evolution of the phase properties, with increasing parametric excitation amplitude, is more suitable for identification purposes.

Just at the stability threshold, the system without any external excitation is neutrally stable, and therefore it may oscillate periodically, so that the response phase has a certain fixed value. In the other extreme case of a purely external random excitation, the phase is distributed uniformly within the interval $[0,2\pi]$; this means that all values from this interval are alike to the phase, in the same way as all places were alike to Kipling's Cat Who Walked by Himself. Now, it is clear that, with increasing parametric excitation amplitude, the transformation from the latter case to the former one, where phase probability density is governed by Dirac delta-functions, should be continuous. And this is indeed the case; however, this continuous transformation of the phase distribution, as the regime of neutral stability is approached, can be observed only if some external excitation does exist, which indeed highlights such an evolution in the nonzero subcritical response.

This evolution is illustrated in Fig. 7.2, where curves of the phase probability density $w(\varphi)$ are presented for several values of the parametric excitation amplitude λ. At $\lambda = 0$, $w(\varphi)$ is constant everywhere (the case of Kipling's Cat). When periodic parametric excitation is present, certain preferable values of phase appear, which coincide with its fixed stable values at the stability threshold $\lambda = \lambda_*$. With increasing relative amplitude $\mu = \lambda/\lambda_*$, these peaks of $w(\varphi)$ become sharper and narrower, and finally, at $\lambda = \lambda_*$, or $\mu = 1$, $w(\varphi)$ degenerates into a pair of Dirac delta-functions. This double-peak shape of $w(\varphi)$ reflects the fact that the frequency of the parametric excitation is twice as high as that of the response. For example, when a periodic axial force excites flexural vibrations in a beam (see Fig. 3.8), the latter will bend in some direction at the instant of maximal compression, whereas at the next such instant — after the excitation cycle is completed — it will bend in the opposite direction.

It is clear from Fig. 7.2 that the system's stability margin, as expressed in terms of the nondimensional parametric excitation amplitude $\mu = \lambda/\lambda_*$, is closely

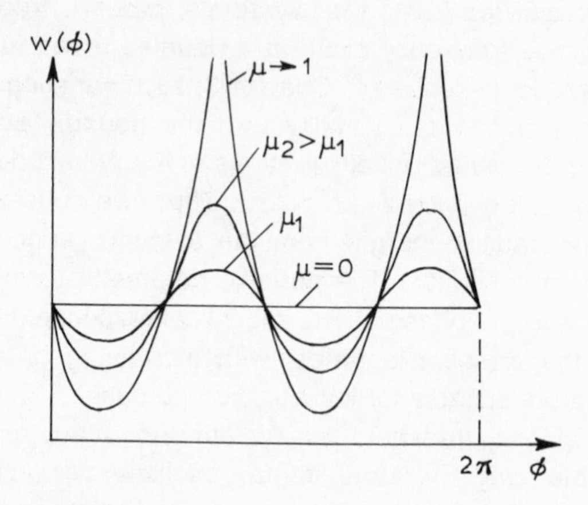

Fig. 7.2 Stationary probability densities of the phase of the narrowband subcritical response of a lightly damped SDOF system to external excitation: evolution with increasing amplitude of the periodic parametric excitation

related to the nonuniformity of the response phase probability density $w(\varphi)$. To be more specific, it has been shown in [21] that this parameter may be estimated from the measured response as a 'phase nonuniformity factor', namely:

$$\mu = \frac{w_{max} - w_{min}}{w_{max} + w_{min}} \tag{7.4}$$

where w_{max} and w_{min} are the maximal and minimal values of $w(\varphi)$, respectively. Moreover, this estimate also simultaneously provides a value for the parametric amplification factor from this formula [21]:

$$\sigma_0^2 = \sigma^2 (1 - \mu^2) \tag{7.5}$$

Here, σ^2 is the averaged-over-the-period variance of the measured response, $x(t)$, whereas σ_0 is the rms value of $x(t)$ where $\lambda = 0$; i.e. in the absence of parametric excitation. Thus the quantity $1/(1 - \mu^2)$ is a parametric amplification factor of the response, whereas its reciprocal $1 - \mu^2$ is a stability margin of an SDOF system with near-resonant parametric excitation. Formula (7.5) provides a direct estimate of the effectiveness of various design modifications, intended, for example, to achieve vibration control by breaking the source of the periodic parametric excitation, or by detuning it from resonance, or by using a vibration absorber. All input data for this estimate can be obtained solely from the measured on-line response signal of the system.

The implementation of this identification procedure, based on estimating $w(\varphi)$, as well as the accuracy of the results obtained, depend strongly on the possibility of measuring a reference periodic parametric excitation signal (the phase is measured, in fact, with respect to this reference). Such measuring should be used whenever possible. For example, if the source of this periodic excitation is clearly identified as being due to the rotation of a certain shaft, measurements of the shaft's rotation angle may be quite useful, even if they are obtained only as discrete marks.

However, computer simulation studies [21] revealed the possibility of estimating the phase probability density, $w(\varphi)$, of a narrowband response $x(t)$, even in the case where a reference signal is not measured and its frequency is not known exactly. The appropriate procedure is to obtain a histogram of the instantaneous frequency $\omega_i = 2\pi/T_i$ of $x(t)$ from its available samples, where the instantaneous period T_i is calculated as the time shift between the ith and i–1th points of the local maxima of $x(t)$. The mode of this histogram, or the most probable value, ν, of the response frequency, was found in these simulations to coincide with one half of the unknown parametric excitation frequency. Therefore, the instantaneous phase $\varphi(t)$ of $x(t)$ may be calculated with respect to that of a harmonic signal with frequency 2ν (using only every second cycle of the latter for phase measurement). More details on these computer simulation studies can be found in [21].

Nevertheless, phase measurements of a narrowband random response may be difficult to implement, since they are highly sensitive to error. Therefore, an alternative approach may be used, which does not require calculations of phase. In this, the inphase and quadrature components $x_c(t)$, $x_s(t)$ of the response are extracted from the available sample of $x(t)$ as described in Chapter 2, where:

$$\begin{aligned} x(t) &= x_c(t)\cos\nu t + x_s(t)\sin\nu t \\ \dot{x}(t) &= \nu\,[-x_c(t)\sin\nu t + x_s(t)\cos\nu t] \end{aligned} \qquad (7.6)$$

Three second-order moments of x_c, x_s are now estimated, namely $K_{cc,ss} = \langle x^2_{c,s}\rangle$, $K_{cs} = \langle x_c x_s\rangle$, and μ is calculated from the formula [21]:

$$\mu = \frac{[(K_{cc} - K_{ss})^2 + 4K_{cs}^2]^{1/2}}{K_{cc} + K_{ss}} \tag{7.7}$$

Thus, an estimation of the phase may be avoided, although an accurate determination of $x_c(t)$, $x_s(t)$ may require many precautions (see Chapter 2). Moreover, this estimate of μ, as well as, for example, the well known estimate of the averaged-over-the-period variance $\sigma^2 = (K_{cc} + K_{ss})/2$, can be shown to be 'robust', or insensitive to possible uncertainties in the known or estimated value of ν. The initial phase of the periodic parametric excitation may also be shown to be irrelevant as long as only the values of μ and σ are sought. Nonzero mean values x_c and/or x_s imply that some external periodic excitation is also present.

This approach has been used in the processing of a vibration signal registered during a 'cold' startup of essentially hydraulic tests on a modified RBMK unit installed at the Ignalina atomic power plant. Spectral analysis of the vibration pickup signal from the exit pipe of a main circulation pump (MCP) revealed several peaks. One of these peaks was very close to one-half of the shaft rotational frequency of the MCP. Although the level of these vibrations was well within safe limits, it was decided to check if this particular peak was related to certain parametric effects, which could imply some anomaly in the shaft operation. Consequently, the signal was passed through a bandpass filter; then its inphase and quadrature components were extracted and their second-order moments calculated. However, formula (7.7) yielded the estimate $\mu \approx 0.06$, which is definitely sufficiently close to zero, within the uncertainty level. Therefore, it was deduced that the observed small vibrational signal was solely due to purely random excitation (presumably a flow-induced one), and was unrelated to the MCP shaft (the rapid decay of the filtered response autocorrelation functions also implies that no external periodic excitation was present at this frequency).

Consider now the same SDOF system for slow, rather than near-resonant, periodic variations of its natural frequency. These variations are assumed to be small, as well as slow, compared with the mean value of the system's natural frequency, so that they cannot lead to an instability. Therefore, the above problems of estimating the system's stability margin and the parametric excitation level are completely irrelevant. However, the detection of such slow periodic natural-frequency variations and an estimation of their amplitude may nevertheless be of some interest for the on-line detection of cracks in a rotating shaft (see Chapter 9).

Analysis of the subcritical response of this system to a broadband random excitation [21] revealed (not quite unexpectedly) that the response amplitude is completely unaffected by these small and slow variations in the system's natural frequency. Moreover, a similar insensitivity is also observed if the one-dimensional probability of the phase is considered, which once again is uniform ($K_{cc} = K_{ss}$, $K_{cs} = 0$). However, the above slow and small variations of the natural frequency influence the correlation functions of the inphase and quadrature components of the response, $x_c(t)$ and $x_s(t)$. In particular, their crosscorrelation function, which is identically zero in the absence of natural-frequency variations, is generally nonzero at nonzero time shifts. In fact, it depends, as do the autocorrelation functions of $x_c(t)$ and $x_s(t)$, on both time arguments, so that these random processes are (periodically) nonstationary. The crosscorrelation function reaches its maximum value with respect to the time shift whenever the latter equals the half-period of the slow parameter variations (where the latter are sinusoidal).

The procedure for detecting natural-frequency variations and estimating their amplitude from on-line response measurement has been verified by computer simulation for sinusoidal variations, with frequency ν [21]. This ratio was particularly suitable for such an estimation:

$$\frac{K_{sc\tau}(t, t+\tau)}{K_{cc\tau}(t, t+\tau)} = \frac{\langle x_s(t)\, x_c(t+\tau)\rangle}{\langle x_c(t)\, x_c(t+\tau)\rangle} \tag{7.8}$$

whereas its values at $t = (\pi/\nu)(k - ½)$, $\tau = \pi/\nu$ $(k = 0, 1, 2, \ldots)$ directly yield an estimate of the amplitude λ of the natural frequency variations:

$$\frac{\lambda\Omega}{\nu} = \arctan \frac{K_{sc\tau}(-\pi/2\nu, \pi/2\nu)}{K_{cc\tau}(-\pi/2\nu, \pi/2\nu)} \tag{7.9}$$

Here, Ω is the mean value of the natural frequency.

The error in the estimates of λ, as recovered by this method from a simulated response x(t), was in the range 5–10%, provided that the available sample of x(t) contained a sufficient number of 'long' periods $2\pi/\nu$ [21]. The success of this identification procedure depends crucially on an exact knowledge of the frequency ν; this, however, is not a problem whenever the variations in the natural frequency are due to the rotation of a certain shaft with a measurable rotation speed. Reference [21] also describes a procedure for estimating the correlation functions of the processes $x_c(t)$, $x_s(t)$ from a single sample of the response x(t) (while in a computer simulation a direct ensemble averaging is possible, it was decided to use those processing methods that may be applied to real response data). This procedure is by no means trivial, since both these processes are nonstationary; it uses the periodic nature of this nonstationarity.

For periodic, rather than random, external excitation, such slow variations in the natural frequency lead to a pair of additional response components with the sum and difference frequencies $\Omega + \nu$ and $\Omega - \nu$. A similar effect is seen in broadband random external excitation: sidebands in the response spectral density may be caused by the natural-frequency variations; i.e. additional peaks at frequencies $\Omega \pm \nu$. However, this is only a second-order effect, whereas the above effect of the crosscorrelation between the inphase and the quadrature components (at nonzero time shifts) is a first-order one, according to theory, and thus should be much more sensitive, whenever it is used for the diagnosis of small-parameter variations.

The next case is again an SDOF system, but with broadband random, rather than periodic, variations in the natural frequency. Here, both the above basic semiqualitative identification problems are of interest, namely estimating the stability margin and the parametric amplification level. In these analyses, various possible definitions of stochastic stability may be used, as outlined in Chapter 3, together with various indexes of the random response level.

In the absence of periodic excitation, no preferable value(s) of phase should exist, so that the phase should be irrelevant for identification (once again, this is Kipling's Cat Who Walked by Himself). The subcritical response amplitude A(t) is, however, affected by random parametric excitation. In particular, the zero-mean part of the squared amplitude, $V_0(t) = V(t) - \langle V(t) \rangle$, where $V = A^2$, has an exponentially decaying correlation function $K_{V_0V_0}(\tau)$ with a decay-rate factor less than that for a purely external excitation, or constant natural frequency [21], for which formula (7.3) holds. On the other hand, the envelope $\bar{K}_{XX}(\tau)$ of the autocorrelation function of the response itself is completely unaffected by random parametric excitation, so that its normalised decay rate is simply α/Ω; i.e. equal to the modal damping ratio (see eqn. (7.2)).

A theoretical analysis [21] gives the following procedure for response-signal processing. Two autocorrelation functions are estimated from the available sample of x(t), i.e. those of x(t) itself ($K_{XX}(\tau)$) and of the zero-mean part $V_0(t) = V(t) - \langle V \rangle$ of its squared amplitude ($K_{V_0V_0}(\tau)$). For the latter estimated function, an exponential decay-rate factor is calculated by a curve fit and denoted as γ_v, whereas for the former the decay rate of its exponential envelope $\bar{K}_{XX}(\tau)$ is calculated and denoted as γ_x. (Both these calculations can once again be made by representing $K_{V_0V_0}(\tau)$ and $\bar{K}_{XX}(\tau)$ in semilogarithmic form and using only the linear parts of the plots to avoid any distortions due to bandpass filtering, whenever it is applied for an MDOF case.)

Now, whenever γ_V is less than $2\gamma_x$, the implication is that some random parametric excitation is present. (The other possible reason is a nonlinearity in the system's restoring force, leading to a broadening of the response spectral density and thus to an overestimation of γ_x. Therefore, the following conclusions and estimates are valid only if the system is indeed linear; the approaches for the detection of such a nonlinearity from on-line random vibration data are described in Chapter 9.) Moreover, the estimated value of $\gamma_V/2$ may be regarded as a mean-square stability margin, whereas the stability margin with respect to probability may be estimated as $\gamma_x/2 + \gamma_V/4$ (of course, in the absence of random parametric excitation, when $\gamma_V = 2\gamma_x$, both these margins are identical and equal to γ_x) [21]. Finally, the quantity $2\gamma_x - \gamma_V$ provides an index of the level of parametric excitation, whereas the ratio $4\gamma_x/(2\gamma_x - \gamma_V)$ provides an estimate of the important parameter δ, i.e. the ratio of the damping factor to the parametric excitation intensity [21]; stability thresholds with respect to probability, and in the mean square, are $\delta = 1$ and $\delta = 2$, respectively.

As for the response amplification problem, if the steady-state mean-square response $\langle x^2 \rangle$ is regarded as an index of the response level, it is possible to define a mean-square response amplification factor as the ratio $\chi = \langle x^2 \rangle / \langle x^2 \rangle_0$, where the subscript zero refers to a purely external excitation. This factor may be expressed in terms of the measured response parameters γ_x, γ_V as $\chi = 2\gamma_x/\gamma_V$ [21].

These correlational methods are valid only if $\delta > 3$, so that the given system is stable with respect to fourth-order moments; otherwise, the squared amplitude V(t) is not stationary in the wide sense because of its infinite variance. This implies that, where $\delta \leq 3$, the finite-sample-length estimates of $K_{V_0V_0}(\tau)$ are not consistent; i.e. they are susceptible to a high statistical scatter, which generally will not decrease with increasing sample length. Furthermore, whenever the given system is unstable in the mean square ($\delta < 2$), the estimates of $K_{XX}(\tau)$, and therefore those of γ_x, also become meaningless. Then the following questions naturally arise: first, how to check independently if the condition $\delta > 3$ is indeed satisfied for a given response process x(t) and therefore the estimates of γ_V, obtained from those of $K_{V_0V_0}(\tau)$, are indeed consistent; and secondly, how to estimate the stability margin and the parametric amplification factor if $\delta \leq 3$.

Both these questions may be answered by an analysis of the stationary probability density w(V) for the subcritical response x(t), which exists whenever the system is stable with respect to probability, i.e. when $\delta > 1$. (The algorithms for the discrimination between such a response and that of a nonlinear system with $\delta < 1$ are presented in Chapter 8.) One can use an estimate of w(V), together with that of the amplitude probability density p(A), to calculate δ from V_m and $w(V_m)$, where $V_m = A_m^2$ is the square of the most probable amplitude; i.e. A_m is the maximum of p(A); the procedure for such calculations is described in [21]. The value of δ may be estimated also from w(V) by a direct curve fit; this can be done most easily in logarithmic coordinates, where:

$$\ln w(V) = -\delta \ln \left(1 + \frac{V}{æ}\right) + \ln C \tag{7.10}$$

Here, æ and C are two other unknown parameters, where æ is proportional to the external excitation.

Whenever the results of a correlational analysis are inadequate ($\delta \leq 3$), the stability margin with respect to probability, $\delta - 1$, may be estimated in this way from measured response data. Moreover, the mean-square response amplification factor may then be estimated as $\chi = (1 - 2/\delta)^{-1}$ (without using γ_x, γ_v). This factor exists, however, only if $\delta > 2$. If $\delta \leq 2$, another response amplification factor may be used; i.e. that based on the square of the most probable amplitude $V_m = A_m^2$; it is $\chi_m = V_m/V_{m0} = \delta/(\delta - ½)$ [21] (where the subscript zero denotes once again a purely external random excitation).

There is, however, yet another direct solution to the 'excitation-sources decomposition' problem, which may be used both for $1 < \delta \leq 3$ and for $\delta > 3$ as an alternative to correlational methods. It can be shown easily from eqn. (7.10) that [21]:

$$\left(\frac{d}{dV}\right) (\ln w) \Big|_{V=0} = -\frac{\delta}{æ} = -\frac{1}{2\sigma_o^2} \tag{7.11}$$

where $\sigma_0^2 = \langle x^2 \rangle_0$; i.e. σ_0 is the rms response of the system without parametric excitation. Thus, σ_0 may be obtained directly from the slope of the tangent to the estimated curve ln w(V) at the origin V = 0. This property has a simple physical interpretation: in the range of vanishingly small response amplitudes, the intensity of parametric excitation is negligibly small, since it is proportional to the square of the response itself. (Therefore, it also holds for periodic parametric excitation, and this way of estimating σ_0 from ln w(V) was verified in the computer simulation [21] referred to previously; the results were only slightly less accurate than those obtained from the phase probability density of the response.) This estimate of σ_0 is valid for any $\delta > 1$. However, it fails to detect mean-square instability; i.e. whether $\delta \leq 2$ or $\delta > 2$ and therefore whether the estimate of the mean-square response $\langle x^2(t) \rangle$ from a finite-length sample of x(t) is meaningful.

The above algorithms for semiqualitative identification may be applied to the response analysis of nuclear-fuel rods in a parallel coolant flow, provided that the linear single-mode model is adequate; some approaches to validating this approach are given in Chapter 8.

The last cases studied are TDOF systems with combinational resonances due to periodic or random excitation. In both cases, external broadband random excitations are assumed to be present, leading to nonzero subcritical responses of the systems. The spectral densities of the modal responses in each case will then have two distinct peaks at the system's natural frequencies. The problem is to estimate the system's stability margin from the measured signal, which may be a weighted sum of these modal responses.

The solution is essentially the same for both periodic, and random, parametric excitation. It is illustrated in Fig. 7.3. The basic idea here is that, in

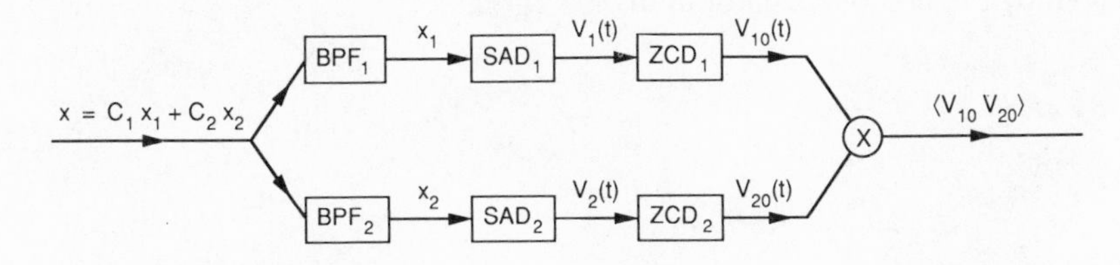

Fig. 7.3 Flow chart of the procedure for estimating the stability margin with respect to the combinational resonance from subcritical response data: $x_i(t)$ — modal responses, BPF — bandpass filters, SAD — detectors of squared amplitude, ZCD — detectors of a zero-mean part of a signal

both cases, the amplitudes of the modal responses become more correlated as the stability threshold is approached. As shown in Fig. 7.3, the measured response signal is first passed simultaneously through a pair of parallel bandpass filters, tuned to its two spectral peaks, with the bandwidth of each filter high compared with that of the corresponding resonant domain. The peaks are assumed to be well separated, and therefore the output signals of the bandpass filters are proportional to the (narrowband) modal responses, $x_1(t)$ and $x_2(t)$. The procedure for extracting the squared amplitude is applied next to the latter, yielding the processes $V_1(t)$, $V_2(t)$. Then, the zero-mean parts, $V_{10}(t)$, $V_{20}(t)$, of the latter are crosscorrelated, and the desired stability margin is estimated as $1-\rho_{12}$, where:

$$\rho_{12} = \frac{\langle V_{10}(t)\, V_{20}(t)\rangle}{\left[\langle V_{10}^2(t)\rangle \langle V_{20}^2(t)\rangle\right]^{1/2}} \tag{7.12}$$

It was shown [21] that, for periodic parametric excitation, $\rho_{12} = 1$ exactly at the stability boundary of the system, whereas for stochastic combinational resonance this equality holds at the stability boundary with respect to fourth-order moments. We get a rather more detailed identification if a time shift between signals V_{10} and V_{20} is introduced for crosscorrelational analysis; i.e. if their crosscorrelation function $R_{12}(\tau) = \langle V_{10}(t)\, V_{20}(t+\tau)\rangle$ is estimated [21]. The absolute maximum $R_{12max}(\tau)$ may be slightly more sensitive to the approach to the stability boundary. Moreover, the sign of the time shift $\tau = \tau_m$, corresponding to this maximum, shows the 'direction of response propagation': if, say, τ_m is positive, then, roughly speaking, the second DOF is driven by the first one, rather than vice versa.

Chapter 8
Discrimination Between Different Types of System from Their On-Line Random Response Data

In this chapter, a further step is made away from complete quantitative identification. Contrary to semiqualitative identification, considered in Chapter 7, we completely abandon any quantitative description of a given unknown system, or 'black box'. The problem is rather to identify, by an appropriate processing of the measured response signals, the type of system (from a given finite set of possible types) producing the responses. Once again, on-line or inservice response measurements, with broadband random input excitation(s) that cannot be measured, are studied.

This is the first problem that faces an engineer whenever a mathematical description of a system with an incompletely known structure is required. Indeed, any attempt at a quantitative identification and/or simulation should be preceded by a check on the adequacy of the adopted mathematical model: if the latter fails to reproduce the principal qualitative features of the observed response, any subsequent quantitative identification, within the framework of the model, may be meaningless. In particular, referring to Fig. 7.1, whenever the system is in a supercritical state ($v > v_*$), its model should be nonlinear; a linear model is adequate (but not necessarily) only if the observed system's response is subcritical ($v < v_*$). (This topic was mentioned in Chapter 7 when the choice of an appropriate model was discussed.) In fact, the systems considered in this chapter are, with few exceptions, nonlinear; therefore, these qualitative identification problems are more difficult (in this respect at least) than quantitative and semiqualitative ones. This is the reason for presenting them after the latter ones, although in general it is the choice of a suitable model that one should begin with whenever one requires an interpretation of measured response(s).

Problems of qualitative identification may arise where prior knowledge of a system is insufficient for an accurate quantitative estimation of its characteristics, whereas simple answers to 'yes or no' or 'either ... or' questions may be adequate in the application. A simple example is that of detecting a periodic signal in a noise background. This problem, which has been studied extensively in radar and sonar applications, may also be of importance for the diagnostics of rotating machinery, such as car engines. Possible applications of qualitative identification may also include the experimental estimation of a system's stability boundaries (parameter v_* in Fig. 7.1) during testing; localisation of a predominant excitation source within a system; etc. Various applications are described later in this chapter.

The first step in qualitative identification is to find a feature of the measured random response signal of the system that can be used to discriminate between systems of various types. The next step is to formulate a criterion for discrimination; i.e. a condition that defines a boundary between the two considered types of system, in terms of their chosen response statistical characteristic. This criterion may uniquely correspond to the introduced classification of the systems. An example is the criterion of a monotonic decrease in the squared-amplitude-response probability for discriminating between forced and self-excited random oscillations. A criterion may be 'biased', as in the monotonic decrease in the probability density of the response itself (within the positive semiaxis) for the same discrimination problem (for a sufficiently high external-excitation-intensity level, the response of a self-oscillatory system may have a response probability density that decreases monotonically within the positive semiaxis). Then, the bias in a theoretically derived criterion should be estimated by computer simulation.

Finally, the third step in qualitative identification is to formulate a decision rule for ascribing a given system to one of the possible types or classes, this decision being based on the appropriate processing of a measured finite sample of the response. The point here is that the theoretical discrimination criterion, on which this decision rule should be based, is usually formulated in terms of certain statistical characteristics of the response, whereas generally, in real life, at best only finite-sample estimates of these characteristics are available. Therefore, some degree of uncertainty will always be present when this discrimination criterion is applied. Moreover, the response data may contain measurement errors.

Therefore, the ultimate check of the identification procedure should be based on computer simulations, in which typical conditions for real-life dynamic data processing are implemented and, possibly, certain extraneous noises and/or errors are introduced artificially. (In view of these requirements, analogue computer simulation is in some respects preferable to digital.)

The first problem is discrimination between forced and self-excited oscillations. Referring to Fig. 7.1, it may be formulated as follows: to determine, by an appropriate processing of the measured system's response during its operation at a certain point of the dashed curve, whether this response corresponds to a subcritical ($v < v_*$) or a supercritical ($v > v_*$) state of the system. In the latter case, the instability of the corresponding linear system is assumed first to be solely due to its self-excitation rather than, say, parametric excitation of some sort. The dynamic instability is assumed here, as elsewhere in this chapter, to be at least partially tolerable for the machine or structure in the sense that corresponding self-excited oscillations do not lead to an immediate failure, so that this supercritical response may indeed be measured.

This problem may be of interest for a mechanical engineer when the stability boundary of a system is to be determined from experiments in different conditions from those in subcritical tests, such as those described in Chapter 7. In many cases, the dynamic instability of a system is only 'mild' with an amplitude growth restricted within safe limits by the internal nonlinearities of the system (these oscillations due to instability may, of course, be undesirable, and 'safe' means here that they lead neither to an immediate failure, nor to significant damage accumulation during the tests). Moreover, this restraint may be provided in the tests by specially designed devices. If during the tests the system is excited by 'natural' random forces, a nonzero steady-state response signal may be registered both in the subcritical (i.e. stable) and in the supercritical (i.e. with an unstable linear part) states. Then some statistical criterion for discriminating between these two states is needed for an accurate evaluation of the stability threshold.

The solution to the above problem may also be useful for the localisation of a predominant source of excitation and for introducing an appropriate design modification for vibration control. For example, violent lateral oscillations of railway vehicles are possible either because of rail irregularities (forced oscillations) or because of vehicle instability (self-excited oscillations). In the first case, the oscillations of the vehicle may be reduced by reducing the input excitation level (e.g. by reducing the tolerances for rail irregularities); whereas in the second case this approach may less effective, and a modification of the system itself (e.g. by changing the parameters of the vehicle suspension and/or the wheel geometry) may be more relevant. Moreover, for many systems the self-oscillatory regime of operation may be known, from field experience, to be undesirable. In such cases, the inservice detection of such a regime from response data will definitely imply that something is wrong with the system.

Among specific mechanical systems for which the above discrimination problem is of interest, many are loaded by nonconservative forces (see Chapter 3). One of the simplest examples is a cantilever beam, loaded by a 'follower' compressive force at one end; i.e. by a force that is always oriented along a deformed neutral axis of the beam. Such a force is produced, for example, by a fluid jet flowing out of a cantilever pipe. It is well known that, above a critical flow speed (and therefore a critical value of the follower compressive force), the straight configuration of a beam or pipe becomes dynamically unstable, and bending vibrations may start. These vibrations will reach a limit, governed by the system's nonlinearities, and eventually a limit cycle with periodic oscillations will be reached. However, in some structures, these vibrations occur at both subcritical and supercritical flows, owing to additional excitation sources (flow pulsations, disturbances due to pumps and/or other extraneous sources etc.), so that discrimination between the sub- and supercritical states is of practical interest. Other systems of this type are: rotating shafts with possible self-oscillations due to internal friction (backward whirl phenomenon [9]), which may also be excited externally through the bearings; elastic structures interacting with a fluid flow, which may vibrate owing to either flutter instability or excitation by a turbulent boundary layer [9]; railway vehicles, which may oscillate laterally both because of their instability and because of excitation from rail irregularities; heat-exchanger-tube arrays in a coolant crossflow, which may vibrate because of vortex-induced excitation and/or because of dynamic instability owing to specific hydrodynamic coupling between adjacent tubes [16].

The solution to the above discrimination problem is based on an analysis of probability densities either of the responses themselves or of their envelopes, or amplitudes. Indeed, the response spectral density is not an appropriate feature for discrimination, since for both the above systems rather similar shapes of spectral density may be observed in most cases (at least, for lightly damped systems); the same can be said about the phase of the response, which in the simplest case of an SDOF system has a uniform distribution for both types of response (forced and self-excited). The clearest case, which is amenable to a thorough theoretical analysis [21], is that of a lightly damped SDOF system with linear restoring force; in this case, the instability is basically due to a 'negative-damping' phenomenon, the growth of the amplitude being restricted only by the (positive) nonlinearity of the damping force. For this case of a single-mode instability, the following criterion has been derived [21], which uses the stationary probability density $w(V)$ of the squared amplitude $V(t)$ of the response $x(t)$.

If $w(V)$ is nonincreasing at every $V \geq 0$, the observed response is a purely forced one; i.e. it is solely due to external random excitation; in the absence of external excitation, no response is observed. This case is illustrated by curve I in Fig. 8.1, and it corresponds to the subcritical range in Fig. 7.1 ($v < v_*$).

If $w(V)$ is an increasing function within a certain interval (V_1, V_2), the observed process represents the system's self-oscillations, perturbed by random excitation(s); this means that in the absence of this(ese) random

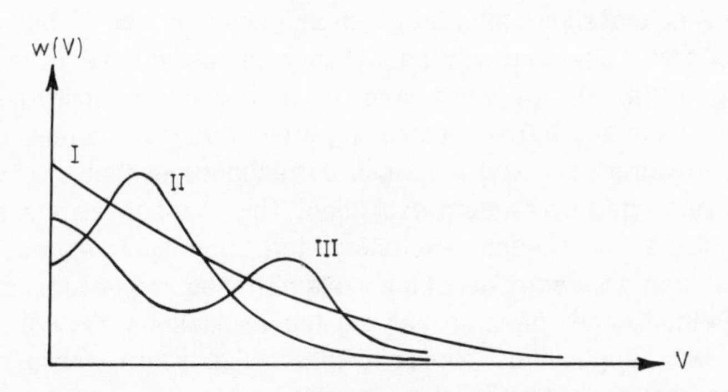

Fig. 8.1 Stationary probability densities w(V) of the squared response amplitude of SDOF systems of various types: I — purely forced oscillations, II and III — systems with soft and hard self-excitation, respectively

excitation(s), the response is periodic. Moreover, if $V_1 = 0$, i.e. if the above interval starts from the origin, the self-excitation is of the 'soft' type, with a static equilibrium position $V = 0$ that is unstable locally. This situation corresponds to the supercritical range in Fig. 7.1 ($v > v_*$) and is illustrated by curve II in Fig. 8.1. On the other hand, if $V_1 > 0$, the self-excitation is of the 'hard' type, i.e. the static equilibrium position $V = 0$ is stable locally, but not globally: sustained periodic oscillations are possible in the absence of input random excitation, and a sufficient initial disturbance transfers the system from the equilibrium state $V = 0$ to this regime of periodic self-oscillations. This case is illustrated by curve III in Fig. 8.1.

This criterion is clear intuitively and is in accordance with this general property of dynamic systems: stable and unstable equilibrium points, limit cycles etc. are reflected in the system's response probability density by the local maxima and minima, respectively, of this density at the above points and cycles. In the above case, this correspondence is exact, irrespective of the random excitation. Therefore, the criterion is 'unbiased', and it provides an exact border between the two types of system in terms of their response to random excitation. Moreover, the maximum V_2 of w(V) in both cases of self-excitation (soft and hard) corresponds to that squared amplitude that would be observed in the absence of random excitation. Thus, a solution may be obtained to the problem of 'excitation-sources decomposition' for this case; i.e. a clear estimate of the oscillation level in the absence of one of the sources (random excitation).

From the above solution, an alternative criterion is derived [21], based on the stationary probability density $w(x)$ of the response itself. Thus, the response is classified as purely forced when it has a bell-shaped or unimodal probability density, monotonically decreasing everywhere at $x > 0$, as in Fig. 1.2(a). On the other hand, for a bimodal probability density $w(x)$, with a minimum at $x = 0$ and increasing within a range of a positive semiaxis (Fig. 1.2(d)), the observed response corresponds to that of a self-excited system, perturbed by random excitation. This criterion also is quite clear. Indeed, for a low random excitation, the subcritical response is near-Gaussian, and a linear model of the system is adequate, at least qualitatively. On the other hand, periodic self-excited oscillations have a 'probability density', with singularities, similar to that in Fig. 1.2(b), and a low random excitation should not destroy this bimodal pattern (although the singularities should disappear); the shape of the response probability density is similar to that for a mixture of a sinusoid with random noise (Fig. 1.2(d)).

This criterion is, however, somewhat 'biased' as it does not provide an exact estimate of the border between the two classes of system in terms of their response probability densities. Thus, for a self-excited system with a sufficiently high random excitation level, the bimodal property of $w(x)$ disappears; in other words, self-excited oscillations cannot be detected by this criterion in a sufficiently high random background. However [21], the 'bias' is not high; i.e. the latter situation applies only to high levels of random excitation. On the other hand, this alternative criterion concerning the shape of $w(x)$ has an important advantage: it does not require the extraction of the amplitude of the observed response signal and therefore may be applied when this signal is not perfectly narrowband; moreover, its implementation within a data-processing procedure is easier, especially in a purely analogue form, without recourse to a digital computer.

Both these criteria have been thoroughly checked in analogue-computer studies [21]. This aspect of the third step in the identification procedure is important: how to establish whether the curve $w(V)$ or $w(x)$, obtained with uncertainty owing to measurement errors and/or the finite length of the available response sample, is monotonically decreasing. This problem is not trivial, since measurements with increments of V (or x) that are too small inevitably lead to a spurious lack of monotonicity due to random errors in w, whereas, with an increment that is too high, the interval with an increase in $w(V)$ or $w(x)$ may be overlooked. In [21], a 'smoothing' procedure is described, based on a sequential pooling of the neighbouring intervals of V or x with too close values of w; it leads to excellent discrimination results, as verified in computer-simulation studies.

Analogue-computer simulation has also been used to verify the second of these criteria ($w(x)$) for a TDOF model of a structure with nonconservative loading. Specifically, the problem was to estimate the stability threshold v_* from random response data where the instability of the linear model is due to a coalescence of the system's natural frequencies, whereas the amplitude growth in a supercritical state is restricted by cubic nonlinearities in the restoring forces ('panel flutter'). The results show that this threshold can be established, according to the $w(x)$ criterion, with an error of about 10%, from the probability densities $w(x_i)$, $i = 1, 2$, of any of the modal responses $x_i(t)$, or from those of their sum.

Therefore, the procedures for discriminating between forced and self-excited oscillations may be applied to many of these systems and structures, loaded by nonconservative forces, to estimate the threshold value v_* of the loading from random response data. One application is a 1:10 scale model of the tree-like RBMK pipeline network, described in Chapter 6 (Fig. 6.3). At the design stage of the RBMK, the possibility of flow-induced vibrations of the pipes was of concern, and this model was built for hydraulic tests. Pipe vibrations were measured, and the authors got tape records of the vibrational signals for correlational and spectral analysis. However, one of their colleagues decided (just for fun!) to apply the discrimination criteria to these signals as well.

The results were as unambiguous as unexpected: the vibrations were self-oscillatory. Now, from the basic laws of mechanics, self-excited oscillations are impossible in a closed system of pipes. A close inspection of the model tests was needed. This revealed the reason for the strange results. In fact, the pipes of the smallest diameter (LWP) were not simulated in the model; instead, just sets of holes of the appropriate diameter were made in the group collectors. Thus, the LWP tubes became free water jets, flowing directly into the surrounding air. They provided a reasonable cold shower, but also a sort of follower loading on the group collector pipes. This nonconservative loading was the source of the dynamic instability, and led to the self-excited oscillations of the group collector pipes that were detected by the analysis of the response probability density. Thus, the proposed criteria clearly showed the need to modify the test model, and their practical value was demonstrated, albeit in a rather strange situation. The appropriate design modifications were introduced: thin flexible rubber pipes were mounted on short tongues with holes on group collector pipes, with the other end of each rubber pipe leading to a large tank filled with water; thus, the free jets were replaced by submerged ones, which did not produce a significant follower loading. The corresponding rms vibration level after this modification was several times less than that of the original model.

The next problem is to discriminate between externally and parametrically excited oscillations where both are random. Assume that the first of the above algorithms has shown the response of an SDOF system to be purely forced; i.e. w(V) is monotonically decreasing everywhere, so that the response is definitely not self-oscillatory. The problem is to determine whether response is due to excitation by an external random force or to random parametric excitation, resulting in a stochastic instability of the system's linear part.

A possible approach to such a quantitative identification problem is to analyse the probability density p(A) of the response amplitude A(t), rather than of its square $V = A^2$. For an external random excitation, p(A) is qualitatively similar to a Rayleigh distribution (see Fig. 1.3(a)), which corresponds to a linear system; in particular, a most probable value A_m of the amplitude exists, at the maximum of p(A). On the other hand, for a purely parametric random excitation, this amplitude probability density may have a different shape. Thus, let the level B_ξ of this excitation be higher than its critical value $B_{\xi*}$, corresponding to an instability threshold in probability (so that the steady-state response indeed exists), but less than $2B_{\xi*}$. Then p(A) is monotonically decreasing everywhere, with a singularity at the origin (solid line in Fig. 8.2). This singularity, however, disappears in the presence of even

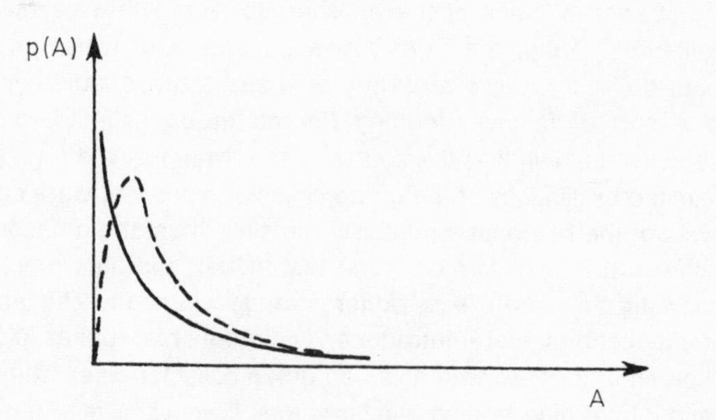

Fig. 8.2 Stationary probability densities p(A) of the response amplitude of a nonlinearly damped SDOF system with random parametric excitation in the presence (dashed line) and in the absence (full line, corresponding to curve I in Fig. 3.11) of random external excitation

a small additional external random excitation, so that the most probable amplitude A_m may be observed; the latter, however, is sufficiently small whenever the level of external excitation is moderate (see dashed curve in Fig. 8.2). Therefore, the observed response is predominantly parametrically excited if $A_m \ll \langle A \rangle$, i.e. if the most probable value of the amplitude is small compared with its mean value.

This criterion, however, is rather 'one-sided' in the sense that the inverse statement is in general not true: if A_m and $\langle A \rangle$ are of the same size, the system does not definitely have a stochastically stable linear part. The first reason is that, for higher external excitation, the most probable amplitude A_m increases and becomes sufficiently close to $\langle A \rangle$, even if the system's linear part is stochastically unstable. The second, more important, reason is that the most probable amplitude A_m occurs even in the absence of external exçitation, provided that the parametric excitation is sufficiently high, i.e. $B_\xi > 2B_{\xi*}$ [21].

In view of this 'bias', an alternative criterion is proposed that is qualitative rather than quantitative. However, it fits the general line of qualitative identification, which is oriented towards a decomposition of the excitation sources. Instead of appealing to the behaviour of p(A) at small A, this approach focuses on the behaviour of the stationary probability density w(V) of the squared amplitude at relatively high values of V. Thus, the random excitation is predominantly parametric when the measured curve ln w(V) is convex, and predominantly external when it is concave. These cases are illustrated by curves 2 and 3, respectively, in Fig. 8.3: whereas the straight

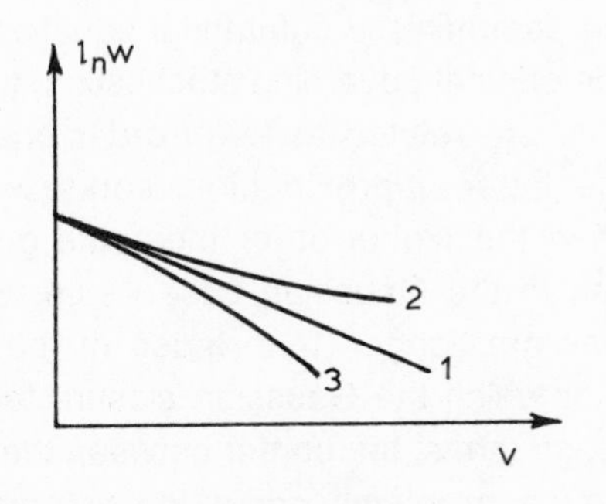

Fig. 8.3 Stationary probability densities of the squared response amplitude of a nonlinearly damped SDOF system in semilog coordinates: 1 — pseudolinear case, 2 — case of a prolonged tail where random parametric excitation prevails, 3 — case where nonlinearity and external excitation prevail

line 1, dividing the above types of excitation, represents the pseudolinear case. The latter corresponds to a Gaussian steady-state response x(t) of the system. (These probability densities are not normalised here, so that the curves of ln w(v) in Fig. 8.3 are shifted arbitrarily along the vertical axis; to illustrate the basic approach in the clearest way, their origins are brought to the same point on the vertical axis).

The rationale behind this criterion is simple [21]. While parametric excitation prolongs the tails of the response probability densities (curve 2), the (positive) nonlinearity of the damping tends to cut them short (curve 3). In the pseudolinear case, these two effects cancel each other out completely, leading to a Gaussian stationary response, similar to that for a linear system with purely external random excitation. Thus, the proposed criterion is natural, as long as the ultimate goal is to estimate, somehow, the influence of the random parametric excitation on the measured overall response. Moreover, these qualitative considerations are substantiated by some rigorous analyses, presented in [21]. Therefore, this approach may indeed be helpful for qualitative identification, although one admits that it is not related directly to the problem of estimating the stochastic stability threshold from random response data; such an estimate may be obtained explicitly, when required, from a direct curve-fit to the measured probability density w(v). By the way, in the linear case, only curves of types 1 and 2 are obtained, implying that parametric amplification of the response may be present (see Chapter 7).

The pseudolinear case plays an important role in statistical dynamics. Thus, a Gaussian closure technique is widely used to obtain solutions to various dynamic problems in nonlinear stochastic systems. This technique permits one to derive deterministic differential equations for the moments of the response from the original governing stochastic equations, assuming that higher-order moments are related to lower-order ones as if the response were Gaussian. The latter approximation works well when nonlinearity prevails, but not when the higher-order moments grow more rapidly with increasing order than in the Gaussian case — i.e. where the tails of the response distributions are 'longer' than those of the Gaussian ones. The pseudolinear case, for which the Gaussian closure techniques yields exact results, defines, in rough terms, the border between these two domains.

Similar discrimination problems occur in systems containing periodic excitations. Once again, it is assumed that broadband random external excitation is also always present, so that for any type of unknown system, or black box, there is a measurable response signal.

Consider first the following problem for an SDOF system with periodic parametric excitation and (positive) damping nonlinearity: to determine, by an appropriate response analysis, whether this system has a stable or an unstable linear part. In terms of Fig. 7.1, this means that we are trying to establish whether there is a subcritical ($v < v_*$) or a supercritical ($v > v_*$) response. In this case, however, v may be regarded as the amplitude of the periodic parametric excitation, and v_* as its threshold value at the stability boundary. The properities of phase response are once again inconvenient for such a discrimination, since the phase probability density is nonuniform in both cases, although in the supercritical case its peaks are sharper.

Therefore, the stationary probability density w(V) of the squared response amplitude is chosen as the appropriate response feature for discrimination purposes. Theoretical analysis and analogue-computer simulations [21] show that the criterion of a monotonic decrease of w(V) is valid, with an established 'bias'. Thus, if w(V) increases within a certain finite range of squared amplitude, the observed response is definitely supercritical. On the other hand, if w(V) decreases monotonically within the whole positive semiaxis, one of two possible conclusions can be made: either the response is subcritical or it is supercritical, but with a high random-amplitude variation. The latter may be defined in terms of the nondimensional ratio $\sigma/\sqrt{V_0}$, where σ is the rms displacement of the corresponding linear system in the absence of periodic parametric excitation, and V_0 is the square of the response amplitude in the absence of external excitation.

Figure 8.4 presents the results of a theoretical analysis (solid curve) and an analogue-computer simulation (circles and triangles) [21] in terms of the 'pseudocritical' parametric-excitation amplitude $\bar{v}_*$. The latter is the value of v

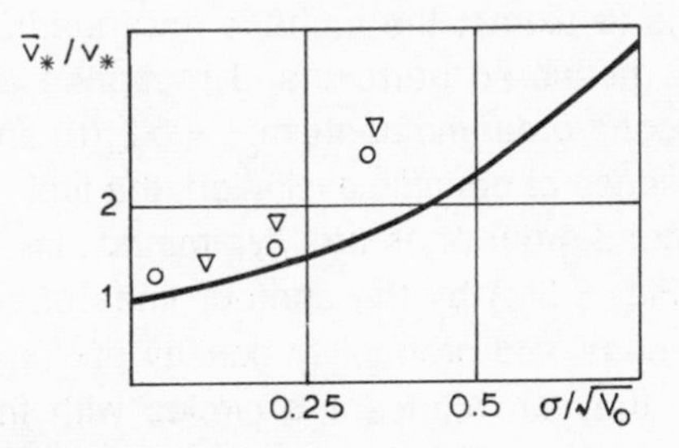

Fig. 8.4 Stability threshold $\bar{v}_*$ of an SDOF system with periodic parametric excitation, as determined from the system's response to random excitation according to the criterion of a monotonous w(V), normalised to the 'true' threshold v_*; ——— theory for the special case of zero detuning, ○, ∇ — computer simulation for zero and nonzero detuning, respectively

such that w(V) monotonically decreases at every V > 0 when v is less than v_* and has a finite range of increase if $v > v_*$. Therefore, it provides an estimate of the 'real' critical value v_* in terms of the measured response. The ratio v_*/v_* is unity at a vanishingly small σ (as it should be, of course). With increasing $\sigma/\sqrt{V_0}$, the monotonicity criterion of w(V) becomes increasingly more 'biased', with a corresponding increase in the possibility of overlooking an instability, because of the high external random excitation. Nevertheless, the criterion is reasonable enough, at least for rough, in-field estimates.

The same criterion may, of course, distinguish such parametrically excited oscillations from purely forced ones, or those excited by external random force only. Then, however, this phase criterion is also efficient: that, for the latter class of systems, the phase of the response is distributed uniformly, whereas for the former it is not. Furthermore, similar criteria may discriminate between SDOF systems with a purely random external excitation and with a combined periodic and random external excitation. The criterion of w(V) monotonicity may detect a periodic excitation component, but once again with a certain 'bias'. Another approach to its detection is to check whether the response phase is distributed uniformly. Finally, when periodic excitation is detected, one can also tell whether it is of an external, or of a parametric, type. The response phase probability density has a single peak within interval $[0,2\pi)$ in the former case, and two peaks in the latter.

Thus, criteria based on the nonuniformity of the phase distribution are quite efficient for the detection of periodic excitation (both external and/or parametric) from the measured inservice response of a system to random excitation. Of course, one faces all the difficulties associated with phase measurement, as described in Chapter 7, when this approach is adopted. An alternative approach is to extract the inphase and quadrature components, $x_c(t)$ and $x_s(t)$, of the measured narrowband response and then calculate their first-order and second-order moments $m_{c,s} = \langle x_{c,s}(t)\rangle$ and $K_{cc,ss} = \langle x^2_{c,s}(t)\rangle$, $K_{cs} = \langle x_c x_s\rangle$. In the absence of periodic excitation, the first-order moments are zero, whereas the second-order ones are 'symmetric', i.e. $K_{cc} = K_{ss}$, $K_{cs} = 0$. This is illustrated in Fig. 8.5(a) by the contour lines of constant stationary Gaussian joint two-dimensional probability density $p(x_c,x_s)$ of x_c and x_s for a linear SDOF system; they are concentric circles with their centre, at the origin, at the maximum of $p(x_c,x_s)$.

When at least one of the calculated expected values $m_c = \langle x_c(t)\rangle$ or $m_s = \langle x_s(t)\rangle$ is nonzero, an additional external periodic excitation may be present. This is illustrated in Fig. 8.5(b) (again for a linear SDOF system). The contour lines are still concentric circles; however, their centre is offset from the origin. The most probable value of the phase is the dashed straight line, drawn from the origin towards the centre. The orientation, or polar angle, of this line, which is the most probable phase angle, is rather arbitrary; it

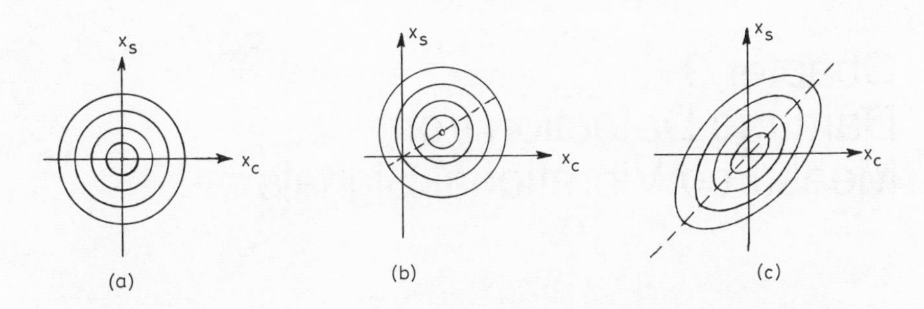

Fig. 8.5 Lines of constant joint probability density of the inphase and quadrature components $x_c(t)$, $x_s(t)$ of the response of an SDOF system to external random excitation: (a) in the absence of any periodic excitation, (b) in the presence of an additional external periodic excitation, (c) in the presence of a periodic parametric excitation

depends on the phase of the reference signal used to extract the inphase and quadrature components of the response. There is only one such most probable phase, corresponding to the first quadrant of the (x_c,x_s) plane in Fig. 8.5(b).

Finally, in the presence of a periodic parametric excitation the second-order moments of x_c, x_s lose their 'symmetry' (for a nonzero μ, where μ is defined by eqn. (7.7)). This is illustrated for a linear SDOF system in Fig.8.5(c). The contour lines of joint probability density $p(x_c,x_s)$ are ellipses rather than circles. The orientation of the common major axis of these ellipses (a dashed straight line) is again arbitrary, and all the ellipses rotate if a different phase of the reference signal is chosen. Two most probable values of the phase now exist, as defined by this line, with a relative phase shift of 180° (one of these corresponds to the first, the other to the third, quadrant of the (x_c,x_s) plane in Fig.8.5(c)). As discussed in Chapter 7, the parameter for second-order-moment 'asymmetry' is governed by eqn. (7.7). Calculation of this parameter for the vibrational signal of the exit pipe of the MCP at the Ignalina atomic power plant (see Chapter 7) provided a qualitative answer: the system contained no periodic parametric excitation.

In this chapter, we provide a formalisation and a quantitative analysis of some of the qualitative dynamic identification problems that face an engineer. The above procedures, as well as more sophisticated ones, are used to analyse the vibrations in nuclear fuel rods, induced by a parallel coolant flow, as described in [21].

Chapter 9
Damage Detection from Measured Vibrational Signals

Anyone who has bought a crystal vase, or a porcelain plate, is familiar with the simple proof test for quality used by the shop assistant, who gently taps the article with a pen or pencil and listens to the sound. When the vase is free of cracks, or similar defects, the sound is quite pleasant. It has a rich content of various modal components, typical generally of musical sounds, and is quite long. In the presence of a crack, quite another type of sound is heard — much shorter, with a higher tone and with fewer modal components. The term 'cracked voice' has its origin, perhaps, in the nature of sounds from cracked solids.

Thus, the shop assistant or any person with suitable experience, may be a good 'diagnostic system' for the detection of cracks in solids and structures, from their noise and/or vibrational signals. The importance of reliable procedures for the detection of cracks and defects or damage, in many machines and structural components, is obvious. We describe now, qualitatively, certain approaches to automatic damage detection using a computer-based diagnostic system that should compete successfully with a human operator. This system, like the shop assistant, works mostly with signals in the audible frequency range, from about 20 to 20000 Hz (in fact, ultrasonic inspection methods have become widespread in various industries, but they have their own limitations and will not be considered here). Moreover, in many cases, on-line detection is required, based on the processing of 'natural' inservice signals in the machine or structure.

The basic difference between ultransonic and vibroacoustic crack detection is that, while ultrasonic wavelengths are small compared with typical crack dimensions (length, depth etc.), the wavelengths of vibrations within the audible frequency range are bigger than the cracks to be detected. This may cause difficulties with detection algorithms in view of diffractional effects (see Chapter 4). Thus, stress waves in a solid with a crack may be only partly reflected by the crack; partly they may circumvent the crack. This may demand sophisticated algorithms for crack detection.

We consider in this Chapter not only simple single cracks, but also other types of damage, such as a broken, or buckled, element of a truss; for brevity, the various types of damage are referred to as 'cracks'. The problem is formulated as follows. Damage within a component reduces its safety and therefore must be detected. If the load-carrying capacity of the component is not exhausted by the damage, the component does not fail immediately, and its noise and/or vibrational signal(s) are measured to detect the damage. Moreover, the size of the crack, or the amount of damage, must also be estimated, to enable one to decide whether the damaged element or component may be left in service or must be replaced. The measured signal(s) may be interpreted using either a standing-wave or a travelling-wave approach, whichever is more appropriate.

Two general types of damage are considered. The first is an open, or 'cutout-type', crack. This involves simple removal of material from part of the component, e.g. when a truss or frame contains a completely broken element. A real crack in a solid behaves like an open one if tensile-stress resultants act at its surface; i.e. when the diagnostic vibrational signal leads to only small dynamic stresses, superimposed on large tensile static ones.

Where there are no comparatively large tensile normal static stresses at the crack surfaces, the crack behaves as a simple cutout within the solid only during those parts of each vibration cycle with tensile-stress resultants at the crack surfaces. When the applied loading leads to compressive normal stress resultants around the crack, its surfaces are pressed against one another; the crack closes, and the behaviour of the cracked solid is the same as that of an uncracked one. Such a crack, which may be either opened or closed, depending on the sign of the time-varying stresses, is a 'flapping' one. This flapping effect may be (and is) used in algorithms for crack detection, as shown later in this chapter.

A simple example is a horizontal beam with a transverse edge crack. In Fig. 9.1, this beam is loaded by a downward static lateral point force, but the same effect occurs for loading by the beam's own weight. This loading leads to bending of the beam so that its lower half is stretched and its upper half is compressed. If the crack is at the lower surface of the beam, it is opened by the downward lateral load, as shown in Fig. 9.1(a) (of course, the displacements of the crack surfaces are exaggerated here). But the loading closes a crack at the beam's upper surface, and the behaviour of a beam with a closed crack is the same as for an uncracked beam; the closed crack is a short line in Fig. 9.1(b).

In Fig. 9.1(a), the lateral deflection of the beam (say, at midspan) is, of course, under the same loading, greater than that of an uncracked one. The reason is that the weakened cross-section with a crack provides additional

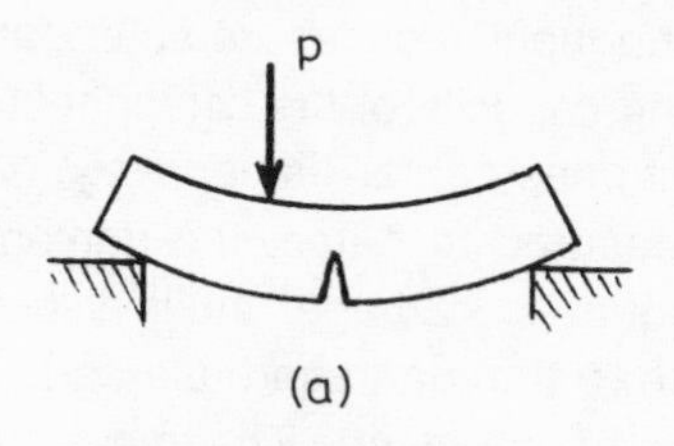

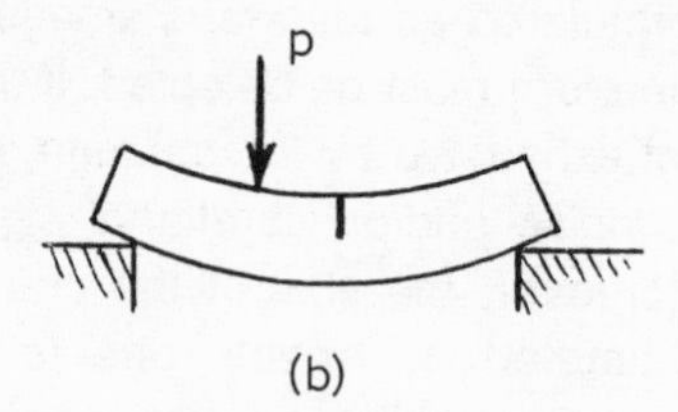

Fig. 9.1 Beam with a transverse edge crack, which may be open whenever it is loaded by tensile stresses (a) and closed when it is compressed (b)

flexibility with respect to lateral loading. In the extreme case of a crack extending throughout the whole cross-section, this flexibility becomes infinite, so that both parts of the cutoff beam deflect separately. By the way, prediction of this reduction in flexural stiffness is quite complex since the simple one-dimensional theory of bending is inadequate in the vicinity of the crack, where the stress and strain fields are inherently two- or three-dimensional. Numerical calculation of these fields, using say finite-element methods, is difficult because of singularities at the crack tip.

In any case, a thin-walled element with open surface crack(s) is roughly an assembly of uncracked parts with appropriate springs at the junction(s). Therefore, deflections of this element may be calculated with simple classical bending theory, as for an uncracked element, so long as the stiffness of the springs, simulating the weakened sections with cracks, is known. For design purposes, this stiffness derives from predictions from solutions to the relevant two- or three-dimensional problems of elasticity, whereas in diagnostic procedures it is estimated from measured vibrational signals. In such models, all the above two- or three-dimensional effects are 'lumped', or localised, within the cracked sections; this may be convenient for diagnostics. For the beam shown in Fig. 9.1(a), the simplest model, with a single rotational spring at the cracked section, is sufficient in the first approximation. Information on the spring constants for such models, for various configurations, can be found in [20].

It is now clear that the relation between a typical displacement x, and the restoring force F, for a cracked element is bilinear, as in Fig. 9.2. Here, the left branch, with a higher slope, corresponds to states with a closed crack. The right branch, with a reduced slope, corresponds to states with an open crack;

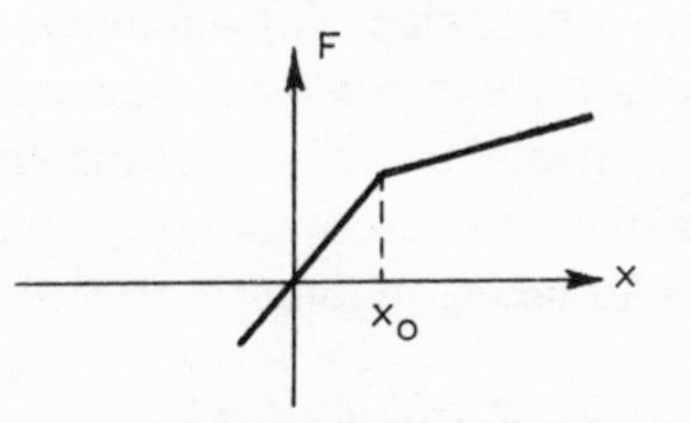

Fig. 9.2 Force-displacement characteristic of an element with a flapping crack; for an unloaded, horizontal beam with a crack at its lower surface, x_0 is the beam's deflection under its own weight at the cracked cross-section

the reduction in the element's stiffness is the result of the additional flexibility of the weakened cross-section. At the break point, with deflection x_0, the resultant force at the crack surface changes its sign. For example, for a heavy unloaded horizontal beam with a crack at its upper surface, x_0 is the beam's deflection, under its own weight, at the cracked cross-section. If the element's displacement $x(t)$ is time-variant, e.g. because of a vibration signal superimposed on static deflection, the crack stays open as long as the displacement stays on the right branch of the bilinear force versus displacement characteristic of Fig. 9.2 ($x(t) > x_0$). However, if it stays on the left branch ($x(t) < x_0$), the crack stays closed and therefore cannot be 'felt' by the vibrational signal. Finally, a vibrational signal with transitions between the two branches of the force versus displacement characteristic means a flapping crack.

Consider first damage as an open crack, or as a completely removed broken element from a truss or frame. Let the available vibrational diagnostic signals lie in a moderate frequency range, where a model with lumped parameters is appropriate for a machine or structure. The diagnostic algorithms are based, here, on measurement of the natural frequencies and/or modal shapes of the structure or component. The results are compared with the reference, or baseline, ones, obtained initially for an undamaged structure (component).

Various aspects of the practical implementation of this approach, such as the choice of the excitation and/or the measured response signal; selection of the natural frequencies and/or the modal shapes as diagnostic indices; etc. are usually case-dependent. Thus, free-vibration tests can usually be made only periodically, during scheduled surveys. But inservice measurement of 'natural' response signals, if any, may permit continuous monitoring of the system's state. The algorithms for extracting modal data (natural frequencies and/or modal shapes) from response measurements are mentioned in Chapter 5. Properly selected modal shapes may be more sensitive to various damages than natural frequencies. For example, fracture or buckling of a stiffening component of a truss or box-type structure, leading to complete loss of stiffness, may mostly affect the local modal shapes of the structure near the failed component, while the corresponding variations in the natural frequencies may be quite small, together with those of certain global modal shapes.

From the measured variations in the spectrum of the natural frequencies of a multielement structure, one may try to identify the element responsible for an observed shift in the spectrum, and the extent of the damage, as expressed in reduced stiffness(es). Such a localisation of damage is, in many cases, impossible, since this inverse problem is ill-conditioned: once again, when quite different sources (damages in this case) lead to similar responses, an accurate identification of the true source (a damaged element, for example) from the measured response(s) may be difficult. Moreover, a particular type of damage may be detected from the measured specific natural frequency only if the latter is sufficiently sensitive to this damage. A continuous tracking (or, at least, periodic measurement) of higher modal shapes of the structure may be preferable. The above considerations should be taken into account whenever a diagnostic measurement strategy is adopted.

Two examples of such diagnostic methods are presented here. The first is for the flexible shaft of a steam turbine in an electric-power plant. In such shafts, lateral surface cracks may develop during operation. One way of detecting the crack is to apply scheduled vibration tests to the shaft after a period of service. The turbine's case is dismantled to remove the shaft, which is then freely suspended horizontally by thin belts or wires. Several of the lowest natural frequencies and modal shapes of the shaft are measured, continuous or transient excitation being provided by a portable shaker. The peak dynamic deflections of the shaft may be smaller than the static ones, owing to the shaft's own weight. Therefore, a lateral crack is open when it is within the lower part of the shaft (see Fig. 9.1(a)). It is essential, therefore, to repeat the tests for various angular positions of the shaft (obtained by rotating it about its own axis), so as to obtain a completely opened crack in at least one of the positions. The variations in the measured natural frequencies and modal shapes, compared with the baseline ones for the reference, or undamaged, shaft, are then used to calculate the depth of the crack, if any, and possibly to estimate its axial position. Approximate flexibilities of a cracked section may be found in [20].

In the above example, a careful continuous monitoring of the system's dynamic characteristics, throughout its service life, is obviously essential. In the second example, continuous monitoring, from the start of the system's service life, is sometimes not so crucial, when even a relatively small damage leads to major changes in the system's dynamic characteristics, particularly its modal shapes. Thus, in systems with high symmetry, many new, additional modes may appear as a result of only a small asymmetric change in the system's properties; these new modes may then be used to detect such changes. A good illustration is a bladed turbine disc: even a small scatter of the blades' masses excites a large variety of additional modes in the disc, with only a small 'mistuning' between the uncoupled bending natural frequencies of the various blades.

An example of the use of this property for diagnostics is an offshore platform [44], consisting of a massive deck, supported above water level by an eight-legged steel truss with a set of bracing elements connecting the various legs. It is important to detect the occurrence of brace severance from measurement of the deck oscillations, without underwater inspection of the truss.

This possibility was studied [44], both by calculating the natural frequencies and modes of the deck's horizontal motions and by measuring them on a 1:50 plastic model of the platform. The results showed that both these types of dynamic characteristic are appropriate for detecting brace severance. However, the modal shapes were much more sensitive to damage than the fundamental natural frequencies: while the latter were reduced by simulated damage by 1 – 4%, the corresponding changes in the normalised modal deck displacement were, for certain modes, 30 – 100%. Moreover, the latter measurement also provided information on the orientation of the damaged component. The measurements can be made directly in the field from a spectral analysis of the response, when measurable oscillations of the platform are excited by waves and/or wind.

The high sensitivity of certain of the deck's modal shapes to damage was probably due to the symmetry of the truss structure, although damage was detected in [44] from drastic changes in the 'old' modes, rather than from the appearance of new ones. Initial measurement of the modes of undamaged structures was thus required in [44]. A continuous tracking of the evolution of the dynamic characteristics, as the basis for diagnostic procedures, may be disadvantageous where inservice variations have other causes. There, special efforts are needed to distinguish these 'natural' variations from the damage-induced ones. Thus, in the offshore platform, a reduction in the deck's fundamental natural frequencies may be due to an increase in deck mass as a result of additional useful loads and/or marine growth on the underwater parts of the structure; an analysis was made in [44] of the consequences of such 'natural' changes so as to distinguish them from damage-induced ones.

A continuous comparison of the system's dynamic characteristics with the original ones, which may be amenable to 'natural' variations irrespective of damage, is not necessary in those procedures for crack detection that are based on new specific effects, owing to the presence of a flapping crack. The basic nonlinear effect is a bilinear restoring force, shown in Fig. 9.2, and the problem is to detect a region with a reduced stiffness in the force versus displacement characteristic from a measured vibrational signal. We start, however, with the special case of a horizontal rotating shaft. Here, the lateral surface crack leads to periodic variations in the shaft stiffness, since it periodically closes and opens under the shaft's weight, as shown in Fig. 9.1. A small vibrational signal does not affect these periodic variations. Therefore, for crack flapping due to shaft rotation, rather than to the vibrational signal itself, the basic effect is of a time-varying, rather than a nonlinear, stiffness.The problem is to detect these periodic variations and to estimate their level on-line, from the measured vibrational signal of the shaft.

The most common type of excitation, which may provide a vibrational signal for diagnostic purposes, is that due to the unbalance of a rotating shaft, i.e. an offset of its mass centre from the geometric centre. The shaft's response to such an excitation depends strongly on the angular orientation of the crack with respect to this unbalance, i.e. to the straight line passing through the above two centres. A detailed analysis of this response to unbalance, with implications for crack detection, is given in [20]. In any case, the development of a lateral crack in a shaft may lead to a continuous growth of the spectral peak corresponding to its response component with the rotational frequency; this criterion, however, may not always be sensitive enough for a timely detection of a crack.

A feature of the response of a horizontal cracked shaft is the 'natural' excitation of the rotating shaft due to its own weight, rather than to unbalance. Thus, while the synchronous response component, with the shaft's rotational frequency, is usually dominant, or may even be the only one, a second-harmonic component, with twice the rotational frequency, may become dominant when the rotational frequency is close to one-half of the shaft's natural frequency. This phenomenon is clearly related to parametric excitation due to a lack of axial symmetry of the shaft's flexural stiffness in the presence of a crack. The resulting second harmonic of the response, with twice the rotational frequency, may resonate with the shaft's natural frequency. This effect is sometimes used to detect a crack during shaft rotation, provided that certain conditions are met, which preclude the growth of the second-harmonic component for other reasons. First, the procedure is applicable only to axisymmetric shafts; otherwise, it may detect the natural shaft asymmetry, rather than a crack. Secondly, various sources of quadratic nonlinearity in the shaft system (for example, in the bearings' fluid films), which may also lead to the growth of a second harmonic, are of minor importance. Furthermore, this effect of the dominance of the second-order harmonic may be observed, if at all, only within a rather narrow range of rotation speeds: the relative width of this range is of the order of the relative stiffness variations due to the crack. In any case, the effect of an increase in the second harmonic with crack growth may indeed be sufficiently sensitive for developing the corresponding diagnostic procedure based on periodic measurement of this increase in service life.

Leaving a more detailed analysis of rotating machinery for the next chapter, we consider now a conceptual procedure for crack detection that may be used when a rotating shaft is excited during its operation at one of its fundamental modes, by some external broadband random forces. In a steam- or gas-turbine shaft, these forces may be aerodynamic. The advantage of the procedure is that it does not depend on the unknown angular orientation of the crack, which is irrelevant.

The response measurements can be made directly on the operating shaft or in special rotation tests. The rotational frequency ν is assumed to be smaller than the natural frequency Ω. This is indeed the case for many steam-turbine shafts with subcritical operational speeds.

The algorithm for detecting relatively slow (subcritical) periodic variations in stiffness is given in Chapter 7 for an SDOF system with broadband external random excitation. It is repeated here briefly in its application to a rotating cracked shaft. The measured shaft response signal is first bandpass filtered, with central frequency Ω, to extract the modal response component at this natural frequency*. From the narrowband output signal $x(t)$ of the filter, the slowly varying random inphase and quadrature components, $x_c(t)$ and $x_s(t)$, are extracted, as described in Chapter 2. The auto- and crosscorrelation functions of the latter are then estimated, particularly near the time shift π/ν, corresponding to one half-period of rotation. A nonzero crosscorrelation between $x_c(t)$ and $x_s(t)$ indicates the presence of the crack, while the above functions at this time shift give the stiffness variations and thus, ultimately, the crack size or depth. An accurate knowledge of the natural frequency Ω is not so crucial for this procedure as that of ν. This may be important, since the rotation speed of shafts can usually be measured with a high precision.

Consider now the nonlinear effect of crack flapping due to transitions in the vibrational signal itself between the two branches of the force versus

---

* This spectral peak can easily be distinguished from that related to the rotation speed, since it shifts only slightly when the latter is varied.

displacement characteristic in Fig. 9.2. The basic case is an SDOF system with such a bilinear restoring force. The diagnostic problem is: from a measured response signal, first, to detect the right branch with a lower slope, when it is present in the restoring force, and secondly, to estimate the difference between the two slopes. This difference is directly related to the crack size or depth.

Estimating the restoring-force nonlinearities from the dynamic response data is also important in other applications. For example, a similar problem arises in vibroimpact systems, when it is of interest to detect the transition into an impact regime of motion and to estimate the impact velocity level*; such estimates are important in assessing the system's reliability. Another application is the detection of backlashes and other assembly faults from response data.

In controlled sinusoidal input excitation of the system, the desired diagnostic algorithms can be based on measurement of the sub- and/or superharmonics of the response signal, which are present only if the restoring force is indeed nonlinear. This idea has been implemented in a system for detecting a cracked spar in an aircraft wing [49]. Vibrations of the wing were excited by a portable shaker, installed on the wing during a postflight inspection. The subharmonic of the order ½, i.e. the response component with one-half of the excitation frequency, was the most sensitive to a restoring-force nonlinearity, and thus most suitable for crack detection [49].

Consider now the same SDOF system for inservice measurements, where the response signal is the result of an uncontrolled 'natural' broadband random excitation. This situation may be visualised in the above example of an aircraft wing with possible cracks in its spar, if an attempt is made, conceptually, to detect cracks in flight from the measured wing response to atmospheric turbulence.

The most obvious approach to this problem of diagnostics is a direct estimate of the stationary probability density $w(x)$ of the response $x(t)$. This density is proportional, in semilog coordinates, to the system's potential energy [19,21], or the integral of the restoring force $F(x)$ (up to an additive constant), i.e.

$$\ln w(x) = \ln C - k \int_{x_0}^{x} F(x')\, dx' \tag{9.1}$$

---

* These systems are governed by specific equations of motion, but on transformation the equations are reduced to those of a system with piecewise–linear restoring forces [21]

(here, C and k are constants, whereas the choice of x_0 is arbitrary: variations in x_0 lead only to vertical shifts of the curve w(x)).

Therefore, a reduction in slope within part of the force versus displacement curve F(x) (corresponding to the right branch in Fig. 9.2) is reflected immediately in the stationary response probability density w(x). Thus, a reduced curvature appears within a part of the semilog plot of w(x), with the same break point as in the F(x) curve. The difference between the two (constant) curvatures provides a direct estimate of the stiffness reduction due to the presence of the crack, and thus of the crack size. This approach to the detection of nonlinearities of the restoring forces on the basis of eqn. (9.1) is applicable to other nonlinearities besides the bilinear one. Its disadvantage is the need for quite a long sample of x(t) to obtain a sufficiently accurate estimate of w(x); small changes in the curvature of the ln w(x) curve are difficult to detect if there is a significant statistical scatter in the estimate.

Two other methods of crack detection are based on the second-order moments only of the measured response, and therefore the requirements concerning the available sample length of the response signal may be less stringent. The price to be paid for this is the more sophisticated algorithms intrinsic in these methods.

The first method, proposed originally in [50], is based on a 'splitting' of the response spectral peak owing to a bilinear nonlinearity. Assume, for simplicity, that an SDOF system is lightly damped, so that each response cycle is close to that of the free vibrations of the corresponding undamped system. Then, each small cycle is sinusoidal, with the natural frequency Ω of the undamaged system. At amplitudes higher than x_0 (see Fig. 9.2), each half cycle with positive x is again almost sinusoidal, with the smaller natural frequency $\Omega_1 < \Omega$ of the damaged system; therefore, the limiting frequency Ω_{min} of such high-amplitude oscillations is $\Omega_{min} = 2/(1/\Omega + 1/\Omega_1) = 2\,\Omega\,\Omega_1/(\Omega + \Omega_1)$.

Such a 'clustering' of the frequencies of the vibration cycles at Ω and Ω_{min} leads to the formation of two resonant peaks in the response spectral density. This splitting of the resonant-spectral-density peak may be used for crack detection if the reduction in the slope, and therefore in the stiffness, due to crack presence is not too small; otherwise, the double peak may be 'smeared', for a system with a finite bandwidth due to damping, or may be undetectable from noise-corrupted measurements. Fig. 9.3 shows a small split in the resonant peak of the response spectral density. This effect may appear in the time domain as a beat phenomenon in the response autocorrelation function (as in Fig. 1.4(d)). Here, the beat period (i.e. the period of the envelope of the response autocorrelation function) provides a direct estimate of the slope difference between the two branches of the force

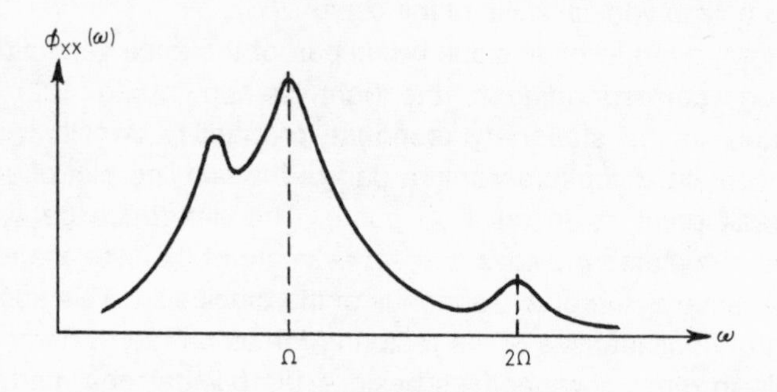

Fig. 9.3 Spectral density of the response of an SDOF system with a bilinear restoring force, such as shown in Fig. 9.2, to a white-noise excitation (two distinct peaks at the main resonance are not always observed)

versus displacement characteristic, and thus of the crack size. Of course, the smaller this difference, and therefore the required frequency resolution, the higher, according to the Uncertainty Principle, the required sample length of the response signal.

The last, but not least, method of crack detection from random response data is based on an analysis of the higher-harmonic components in the response signal. In fact, any response cycle containing transition(s) between two branches of a force versus displacement characteristic (Fig. 9.2) is definitely nonsinusoidal. Therefore, it contains higher harmonics, which accompany the fundamental one. A sample of the response of a lightly damped SDOF system to random excitation is, roughly, a collection of such nonsinusoidal cycles of free vibration with a random amplitude of each cycle. The higher harmonics are reflected in additional peaks in the response spectral density, at various integer multiples of the undamaged system's natural frequency. Of course, for a small crack, and consequently for a weaker higher-harmonic content of the response, these peaks may appear as small 'humps' on the background of the nonresonant part of the response spectral density. Such a hump is shown in Fig. 9.3 at twice the system's natural frequency. This method is applicable to any type of nonlinearity (contrary to the previous method, based on a resonant-spectral-peak split).

The problem is reduced, therefore, to that of detecting additional peak(s) in the response spectral density at multiple(s) of the natural frequency of the undamaged system. Moreover, the level(s) of this(ese) peak(s) is (are) directly related to the level of apparent system nonlinearity (a stiffness reduction for a crack) and may be an index of crack size or depth. However, direct identification of such a peak may be difficult because of the inherent background, or 'pedestal', owing to the nonresonant-response component at the frequency of the peak. Other modal response components and/or measurement errors may also contribute to the masking of this(ese) additional spectral peak(s).

A way to overcome this difficulty is to use a crosscorrelation analysis between the amplitudes of the response components with frequencies near the resonant frequency and its multiples. The amplitudes of the different frequency components of a zero-mean stationary random process are uncorrelated. Therefore, nonzero crosscorrelation is caused only by an inherent coupling of the higher-order response harmonics with the fundamental one, through the common amplitude of each nonsinusoidal response cycle*. Thus, the method is based on the detection of nonsinusoidal cycles, present in the response signal because of the effect of a flapping crack.

The identification algorithm is as follows. The response signal is passed simultaneously through a pair of parallel bandpass filters with identical relative bandwidths. One of these filters has a central frequency Ω; the other is tuned to one of its multiples, e.g. 2Ω or 3Ω (these additional spectral peaks at the lowest-order harmonics are usually the most pronounced ones). The amplitudes of the output signals of both the filters are then extracted, and the crosscorrelation factor of their zero-mean parts is estimated (for coinciding time instants). A significant nonzero value of this factor indicates that a crack is present. The size of the crack can be estimated from its ratio to the product of the rms amplitudes of the two filtered signals; i.e. from the normalised crosscorrelation factor. The latter is less than unity mostly because of the above-mentioned nonresonant 'pedestal' of the higher-harmonic peak if other modal responses and measurement noises are negligible. This pedestal is estimated by assuming that the attenuation rate of the response spectral density with increasing frequency is approximately the same as for the linear system. As for the choice of the filters' relative bandwidth Δ/Ω ($= 2\Delta/2\Omega = 3\Delta/3\Omega$ etc.), it should generally be several times higher than the damping ratio α/Ω, in order to make the procedure more robust (less sensitive) to possible inaccurate tuning of the filters. On the other hand, increasing Δ leads to a reduction in the crosscorrelation between the amplitudes, because of the increased content of nonresonant components, and thus to a reduction in the resolution of the method. Therefore, some tradeoff is required; good results were obtained in a computer simulations with $\alpha = 10\Delta$.

These three methods were verified and compared in many computer-

---

* Crosscorrelation with other modal responses may also be neglected.

simulation runs. All three methods are reliable for a sufficiently high nonlinearity. The first one, however, involves more stringent requirements concerning the available sample length of the response. The method based on a resonant-spectral-peak split does not work when the differences between the small-amplitude and the high-amplitude natural frequencies Ω and Ω_1 are less than about one-third of the former (depending mostly on the damping ratio). Then, a break in the force versus displacement characteristic owing to the presence of the crack may lead only to a broadening and a variation in the shape of the response spectral peak: this peak may become 'smeared', as in a 'usual', or continuous, nonlinearity. For damping ratios higher than 0.5%, this method does not work.

The method of internal crosscorrelation of the amplitudes is much more reliable for a small nonlinearity. The detection of higher spectral peaks by this method is quite consistent, at least in noisefree computer-simulation runs. The method also demonstrates an excellent performance for another type of nonlinearity, i.e. for an SDOF vibroimpact system with a one-sided barrier. In this case, the impacts are readily detected from an additional spectral peak at the threefold natural frequency, using an internal crosscorrelation of the amplitudes.

A conceptual algorithm is now outlined for crack detection in thin-walled structures, such as beams, plates and shells, from their travelling-wave responses. Consider, for simplicity, a beam with a lateral surface crack within a cross-section $x = x_0$, where x is a longitudinal coordinate. The crack is assumed to be open, from sufficiently high static loads; as in a pipeline containing pressurised fluid.

Let the axial and/or flexural one-dimensional waves propagate in the beam, say, 'from left to right', with their wavelengths much higher than the beam's thickness (this is usual when only 'natural' structural vibrations, rather than special test signals, are used for diagnostics). At $x = x_0$, the stress state in the beam is not one-dimensional, since the cracked part of the cross-section is free of stresses, whereas the undamaged part freely transmits the stress waves. However, two- or three-dimensional effects decay quite rapidly as the distance from the cracked cross-section increases. Thus, axial and flexural waves dominate both in the reflected and in the transmitted waves at distances higher than several thicknesses.

Therefore, crosscorrelational methods for detecting reflected and transmitted waves, as described in Chapter 4, can be used to detect possible cracks. For this, one installs vibration sensors near to the crucial parts of the structure, such as the most stressed points, welds etc. Of course, this simple approach works only if the basic response signals do not contain reflected waves from other sources, such as supports, attached masses etc., or, at least, if these 'natural' reflected waves can be identified properly and filtered out. The analysis is done separately, in different frequency bands, since the reflected and transmitted waves always contain flexural components (and are both dispersive), even if the incident wave is purely axial and no crosscoupling between the axial and flexural deformations occurs in the (straight) beam; such a coupling stems from two- and/or three-dimensional effects near the cross-section containing the crack.

Chapter 10
Diagnostics of Rotating Machinery

By rotating machinery, we usually mean machines and/or mechanisms with rotating parts, namely shafts, or rotors, with their bearings, and possibly components such as disc couplings, wheels, cooling fans etc., fixed to the rotating shafts. Such machines are used in practically every industry, and they differ significantly in size; compare, for example, the huge shafts of hydraulic or steam turbines with the small ones used in precision devices such as gyros. Thus, machines with rotating components are extremely important and must be considered in a separate chapter. Another reason to consider the diagnostics of rotating machinery separately is that a measurable noise or vibrational signal usually contains strong frequency component(s) corresponding to the shaft rotation speed and possibly its multiples or harmonics; other frequency components occur in some rotational machines, such as tooth-meshing frequencies in gearboxes or the cage rotational frequency in ball-bearings, to be explained later. The rotation speed in steady operation is usually controlled quite accurately. Therefore, it is quite natural that the fixed rotational frequency(ies) components and certain of its (their) multiples are the most convenient for diagnostics; or, at least, that the existence of a baseline signal component with the rotational frequency cannot but be accounted for and used in some way in most diagnostic algorithms.

Thus, vibrational signals from rotating machinery, which are measured most easily by accelerometers fixed to the bearing cases or supports, usually have lined spectra, consisting of discrete set(s) of various frequency components (of course, a continuous broadband part of the response-signal spectrum can be observed in many cases, as excited by some external random loading, aerodynamic or magnetic forces etc.). These frequency components can be effectively used as signatures of the system's internal state, provided that they are identified properly. The problem of classifying these frequency components can be far from trivial, particularly when several overlapping sets of the components are present, corresponding to the characteristic rotational speeds of the machine. Thus, while in most other machines or structures the interpretation of vibration data may be difficult because of too poor a spectral-peak content in the measured signals, in rotational machinery the problem may be just the opposite: because of too rich a spectral-peaks content, one must identify each of them properly, and sort them out. In any event, a knowledge of the basic reference or characteristic frequencies gives the most thorough interpretation of the measured signal(s), and better information on the system's internal state. This is one reason why the diagnostic procedures for rotating equipment are more advanced than those for most other types of machine, mechanism and structure.

To interpret the frequency components of the recorded response signal(s) properly, we must establish their possible sources. One of the main sources of low–frequency components is shaft unbalance, leading to vectorial centrifugal forces and moments, F_c and M_c, respectively. The latter are:

$$F_c = \sum_{i=1}^{n} m_i r_i \Omega^2 = m_s e_{st} \Omega^2 \qquad (10.1)$$

$$M_c = \sum_{i=1}^{n} m_i [l_i r_i] \Omega^2 \qquad (10.2)$$

where the subscript i corresponds to the number of an unbalanced shaft's element; n is the total number of these elements; m_i, r_i are the mass of the i-th element and the offset of its mass centre (MC) from the rotation axis, respectively; l_i is the distance of the i-th MC from one of the two supports; whereas

$$e_{st} = \sum_{i=1}^{n} m_i \frac{r_i}{m_s}$$

is the eccentricity of the shaft with total mass m_s; in a single element, it is simply the offset between MC and the geometric rotor centre (GC). Shaft unbalance can be described, therefore, by the vectorial, static unbalance $D_s = m_s e_{st}$ and the principal moment of unbalance $T_0 = [D_M L]$ (vectorial product), with L the arm (distance between the two supports) of the couple unbalances D_M.

The case $F_c \neq 0$, $M_c = 0$ is observed when the rotation axis and the principal central inertia axis are parallel but not coaxial. With linear misalignment, all the unbalanced masses are reduced to a single equivalent apparent unbalanced mass, and the responses at the supports are the same, both in their magnitude and in their phase. Static unbalance is defined completely by the shaft's eccentricity e_{st} (the shift between the axes 00 and 0′0′ in Fig. 10.1).

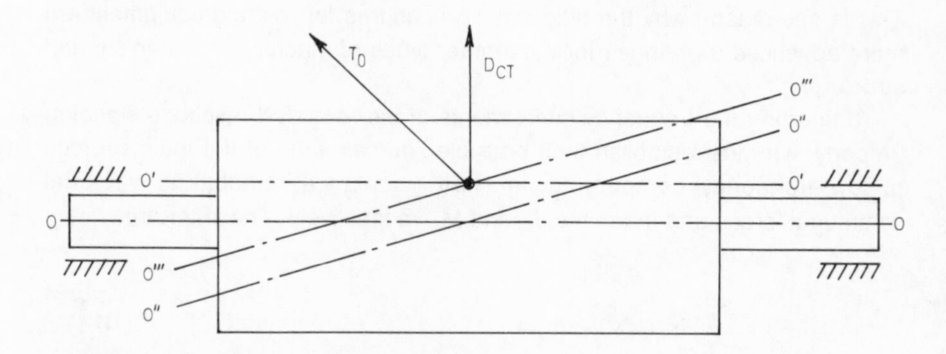

Fig. 10.1 Types of shaft imbalance for various positions and orientations of the main central inertia axis

Angular misalignment of the shaft and/or its fixed components (discs, wheels etc.) may also lead to moment unbalance, with the principal central inertia axis inclined to the rotation axis symmetrically with respect to the supports (axis 0″0″ in Fig. 10.1). Then, $F_c = 0$, $M_c \neq 0$, and all the unbalanced masses are equivalent to a pair of antisymmetric (with respect to shaft MC) masses fixed to the same axial plane within a rotating shaft. The responses of the shaft at the supports are the same, but have a 180° relative phase shift.

In a general misalignment, both F_c and M_c are nonzero, and the corresponding responses at the supports are different. However, on comparing these responses, one can establish, for example, what type of unbalance prevails in a shaft. The procedures for rotor balancing are often based on identification methods.

Of course, vibration of a rotating shaft depends not only on its unbalance, but also on, for example, the shaft and/or support flexibility. The latter governs the critical rotation speeds, coinciding with the natural frequencies of the shaft-support system. In many shaft vibrations, the bending stiffness or flexibility is most important, whereas the supports are roughly quite rigid; as for steam-turbine shafts in electric-power plants. Often, however, the support flexibility is comparable with that of the shaft; as for some aircraft-compressor and/or gas-turbine shafts. Depending on the operating rotation speed Ω, shafts may be classified provisionally as rigid if $\Omega < 0.7\,\Omega_{cr1}$ and flexible if $\Omega > \Omega_{cr1}$, where Ω_{cr1} is the lowest or fundamental critical rotation speed.

Unbalance is the main source of synchronous lateral oscillation, or forward whirl, of a shaft, i.e. of a dynamic displacement pattern that rotates with the shaft (so that an observer, placed on the rotating unbalanced shaft, sees the lateral displacement field due to unbalance as completely 'frozen'). A backward whirl or precession may be excited in a shaft owing to its instability. A source of the instability may be the 'internal' damping forces within the shaft, which rotate with it. These may be destabilising [9,18], contrary to 'external' damping (such as that within supports), which is always stabilising. Furthermore, torsional oscillations may also be excited in shaft systems, by time-variant torques of the drive and/or the load.

Parametric-excitation phenomena may occur in a rotating shaft when its cross-section is not axisymmetric (e.g. owing to a crack or a key-groove), so that its stiffness in a fixed plane depends on the rotation angle. This may greatly increase the amplitude of the second rotational frequency harmonic, relative to that of the main, or synchronous one, when the rotation speed equals one-half of the first critical one, as explained in Chapter 9. Periodic parametric excitation may also come from a gear mesh, with inherent periodic variation in the number of pairs of teeth in simultaneous contact. In spur gears, either one or two pairs of teeth may be in contact simultaneously, so that the mesh stiffness exhibits twofold variations, their frequency being equal to the meshing frequency; in helical gears, these periodic variations are not so drastic, and this explains their lower noise.

Parametric instability is not observed in properly designed machines and/or mechanisms with rotating components. Parametric excitation manifests itself, if at all, in changes in the system's response to external excitation. It leads to additional frequency components, or sidebands, in the response spectrum. For example, in a periodic external excitation with frequency ν, periodic variations in an SDOF system's natural frequency with frequency $\omega < \nu$ lead to additional response components (sidebands) with frequencies $\nu + \omega$ and $\nu - \omega$. This also happens in broadband random external excitation, with corresponding peaks in the response spectral density; however, it is of minor importance compared with the effect of crosscorrelation between the inphase and quadrature components of the response, which can be used to identify slow periodic variations in the system's natural frequency (see Chapter 9). In parametric resonance, these variations may lead to some amplification of the (subcritical) response to external excitation. This amplification can be estimated for measured responses by the methods described in Chapter 7.

Important sources of vibration in rotating mechanisms of small and medium size are their rolling-element bearings. A feature of these sources is the additional harmonic-frequency components with various characteristic frequency(ies) of the bearings. Since some of these components are present in the measured response signals, it is important to correlate the measured response peaks in the frequency domain with the above characteristic frequencies; in fact, it is the first step in response-data interpretation for shafts with rolling-element bearings. The characteristic frequencies are calculated as follows.

Figure 10.2 shows a radial ball bearing with doubly curved races; the outer

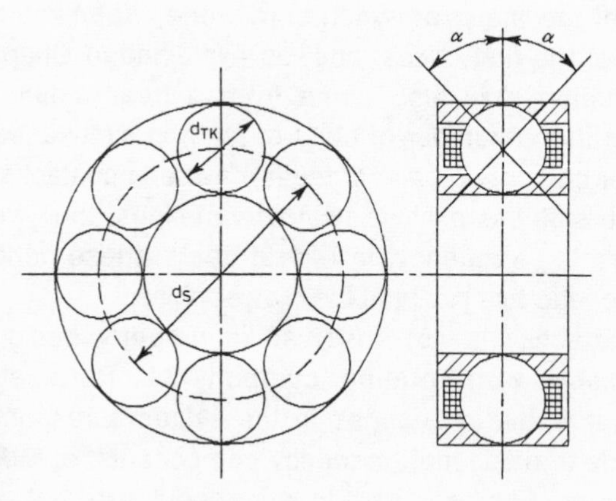

Fig. 10.2 Sketch of a ball-bearing

race is fixed whereas the inner one rotates with the shaft. Here, α is the contact angle, with values up to 36°; it provides some axial-thrust-load capacity. Often, though, this angle is small, and one may put $\cos\alpha \approx 1$ in the following formulae, as well as in the case of roller bearing with cylindrical rather than spherical rolling elements. The relations between the characteristic frequencies are derived simply from the condition for pure rolling; i.e. of equal linear velocity of the two contacting bodies at their point of contact.

For an inner race fixed to the shaft (as in Fig. 10.2), rotating with frequency ω_s, the rotational frequency of the cage (ball carrier) ω_c is:

$$\omega_c = (\frac{\omega_s}{2})\,(1 - d_{re} \cos \frac{\alpha}{d_c}) \tag{10.3}$$

Here, d_{re} is the diameter of the rolling elements, whereas d_c is the cage diameter, defined as the diameter of the circle passing through the rolling-element centres. The rolling element itself rotates around its centre with frequency:

$$\omega_{re} = \omega_c\,(\frac{d_c}{d_{re}} + \cos \alpha) \tag{10.4}$$

For diagnostic purposes, the frequency, or rate, of passage of the rolling elements along the inner and outer races is important. The second of these, ω_o, is, in fact, the product of the cage rotational frequency and the number of rolling elements z_{re}:

$$\omega_o = z_{re} \,.\, \omega_c = (\frac{\omega_s}{2})\,(1 - \frac{d_{re}}{d_c} \cos \alpha) \,.\, z_{re} \tag{10.5}$$

The stiffness of a rolling-element bearing depends on the angular position of the rolling elements, and therefore varies periodically with the above frequency ω_o. Thus, rolling-element bearings may be a source of parametric excitation in shaft systems.

Similarly, the frequency of passage of the rolling elements along the inner race is:

$$\omega_i = (\omega_s - \omega_c)\, z_{re} = (\frac{\omega_s}{2})\,(1 + \frac{d_{re}}{d_c} \cos \alpha)\, z_{re} \tag{10.6}$$

This frequency is observed in a response signal from a preloaded bearing. Various multiples and combinations of the above frequencies occur in the shaft response spectra. Shaft vibrations depend significantly on machining errors in rolling elements, such as out-of-roundness or scatter of sizes. The first factor may be different for shafts with ball and cylindrical roller bearings.

Many mechanical vibration sources are related to machining and/or assembly defects in rotary machines, as well as to inservice wear. One common shaft defect, which may induce bearing vibration, is out-of-roundness of its journal(s), leading to a kinematic excitation of the shaft at twice its rotation speed, with constant amplitude. Similar vibration is induced by such defects of the bearing inner race. Quite common defects are shaft unbalance and/or asymmetry, as well as misalignment of shaft(s) and/or discs, couplings, bearings etc. [53].

As already stated, the amplitudes of the periodic frequency components of a shaft system's response are used as diagnostic indices. The selected indices should be those most sensitive to expected defects or faults and least sensitive to variations in environmental conditions and operational regimes. Thus, shaft unbalance and asymmetry of its stiffness may be detected best of all from its response level at the shaft rotational frequency and possibly at its low-order multiples.

A complex pattern of response spectra occurs in shaft systems with rolling-element bearings; sophisticated algorithms may be required to sort out the various sets of frequency components for bearing diagnostics. For example, suppose that some event with a rotational frequency is present within the system, such as that due to periodic impacts. If the corresponding response components at the rotational frequency, and its lowest-order multiples, are considered, the effect of this event may be masked by those due to shaft imbalance and/or misalignment. The latter, however, may be rapidly attenuated for higher-order multiples, whereas the response to impacts may persist in the frequency domain up to very high frequencies. This suggests the following diagnostic procedure for the case where the above 'pulse train' excites measurable resonant oscillations at a sufficiently high natural frequency of the rotor-bearing system. The response is bandpass-filtered, so that only near-resonant components, around this natural frequency, are retained. Then the envelope, or amplitude, of the filtered signal is extracted. The latter operation implies a shift of the signal spectrum to the origin in the frequency domain.

This procedure is often used for rolling-element bearings [1,34], with a spectral analysis of the response amplitude, or envelope, providing useful information on the various types of fault. Indeed, every type of fault or defect should show up as a harmonic contribution to this envelope, its specific frequency being related to the basic bearing rotational frequencies in eqns. (10.3) to (10.6) (those of the shaft, cage and rolling elements, i.e. ω_s, ω_c, and ω_{re}, as well as the rolling frequencies along the outer and inner races ω_o, ω_i). Thus, a harmonic component in the envelope spectrum is used to detect and identify the corresponding type of fault, while the amplitudes of the harmonic components, or the partial modulation factors, indicate the extent of the damage.

Consider a system with a horizontal shaft, supported by radial ball bearings. Without assembly errors and wear, the response-envelope spectrum usually contains only one small harmonic component, at the outer-race rolling frequency or the passage rate of the balls along the outer race $\omega_o = \omega_c z_{re}$, defined by eqn. (10.5). A typical response-envelope spectrum of a 'normal' ball bearing (without detectable faults) is presented in Fig. 10.3(a), reproduced from [1]; the rotational circular frequency here is $f_s = \omega_s/2\pi = 50$ Hz.

Assembly errors of the bearing system lead to strong components in the response-envelope spectrum with frequencies $2\omega_i$ and $2\omega_o$, for angular misalignment of the inner and outer races, respectively. These spectra are shown in Figs. 10.3(b) and 10.3(c), respectively; and the reason for the additional peaks in both cases is the appearance of pairs of points with excessive loads on the corresponding misaligned races. Furthermore, transportation of a machine with such assembly defects of the rolling-element bearing(s) may lead to a highly localised strain hardening of the bearing; this is an increase in the stiffness owing to local plastic deformation of the metal under increased local static loads. This local increase in stiffness may lead to periodic impacts with rate ω_o and its multiples, as shown in Fig. 10.3(d).

Angular misalignment(s) of the axes within a rotary machine and its drive may lead to the appearance of a set of harmonic response components at the shaft rotational frequency ω_s and its multiples, the former being most prominent, as shown in Fig. 10.3(e).

Excessive wear in a rolling-element bearing may have its own diagnostic indices, namely modulation factors of high-frequency resonant vibrations at specific frequencies for various types of corresponding defect. Thus, wear of the outer race is accompanied by an increase in the response-envelope-spectrum level, at frequency ω_o, whereas surface cracks, spalls, pits etc. give an increase at multiples of ω_o. Wear of the inner race is accompanied by an increase in the spectrum level at multiples of ω_s; while cracks, spalls and pits generate additional frequency components in the envelope with the passage rate of the rolling elements along the inner race, ω_i. Micropitting of the rolling elements and their nonuniform wear generate additional harmonic components in the envelope with frequencies ω_c and ω_{re}, respectively. With high wear, new, 'combinational', frequency components appear in the response envelope (their frequencies being linear combinations of the basic ones), such as that with frequency $\omega_o + \omega_s$ in Fig. 10.3(f).

Finally, Fig. 10.3(g) illustrates the system's response to yet another fault, namely shaft-neck out-of-roundedness due to manufacturing errors and/or inservice wear. Shaft out-of-roundedness may be an oval or polygonal shape

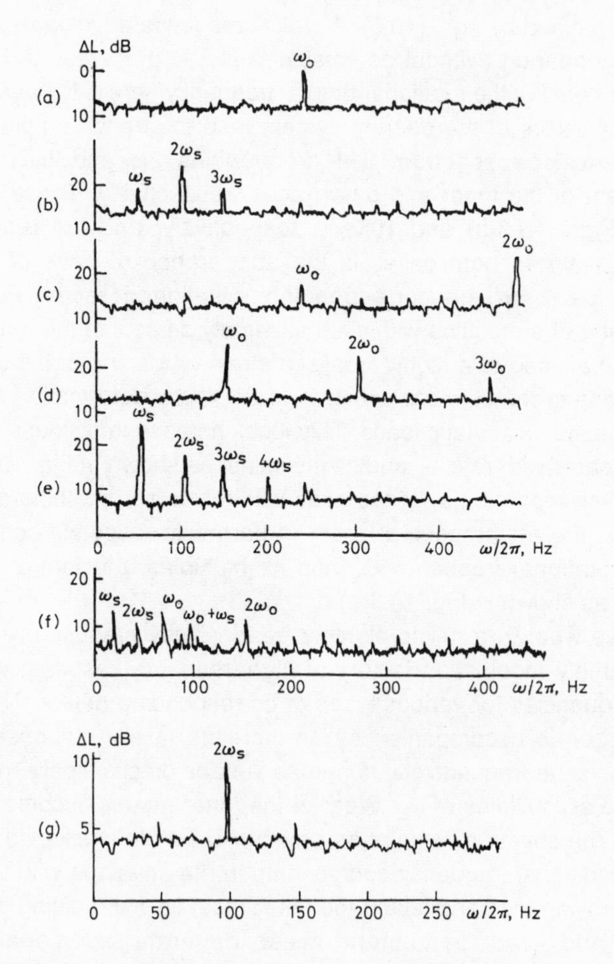

Fig. 10.3 Envelope spectra of ball-bearing case response: (a) bearing without defects, (b) with angular misalignment of the inner race, (c) with angular misalignment of the outer race, (d) with strain hardening, (e) with angular misalignment of the couplings, (f) with high wear, (g) with shaft-neck out-of-roundness (in a system with journal bearings). From [1]

of the neck cross-section. In the first case, shown in Fig. 10.3(g), the high-frequency shaft response at a natural frequency is modulated by a harmonic signal with frequency $2\omega_s$. (In fact, contrary to the other Figs. 10.3, this one is for a shaft with a plain journal bearing, rather than a ball bearing; it is not surprising, therefore, that this response-envelope spectrum is much more uniform at other frequencies; this fact, however, is not of major importance, since Fig. 10.3(g) illustrates the diagnostics of the shaft itself rather than of its bearings.) For a polygonal shaft neck, the corresponding modulation frequency is $\kappa\Omega_s$, where κ is the number of facets.

Often, for the state monitoring of rolling-element bearings, less sophisticated algorithms are used, which may be completely automated and implemented both in computer-based systems and in specific electronic devices. However, these algorithms are less sensitive than a spectral analysis of the response envelope.

Thus, the devices MERA-10A, MERA-21A provide fault detection in bearings on the basis of the SPM-Shock Pulse Method [3]. The latter treats a rolling-element bearing as a random pulse generator. Each pulse is usually small in itself; the device counts the pulses and warns of a fault when the number of pulses in a given time exceeds a preassigned threshold value.

Another single diagnostic index, widely used for rolling-element bearings, is the kurtosis factor æ of the response [34,45] (see the definition in Chapter 1). This factor can be calculated either for the overall response signal, or for its components within different frequency ranges; e.g. these four bands are sometimes used: 3 Hz – 5 Hz, 5 Hz – 10 kHz, 10 kHz – 15 kHz, 15 kHz – 20 kHz [34]. The kurtosis factor usually indicates the closeness of the given signal to a normal, or Gaussian one, since for the latter æ = 0.

The kurtosis is often (but not necessarily) positive for a probability-density curve with a sharper apex, and negative for a less sharp apex, compared with that of a Gaussian curve. The response signal in the absence of significant faults is usually close to Gaussian, with æ ≈ 0. This is to be expected for a signal obtained as the sum of a large number of relatively small signals from various independent sources (see Chapter 1), with comparable levels. Roughly speaking, in this case there are no defects making one of the component signals dominant. Early damage usually leads to variations in the kurtosis factor in the low-frequency band, whereas the subsequent progress of the damage may influence the response within the higher-frequency bands and lead to a return to the initial value of the kurtosis, at low frequencies.

Vibrational diagnostics is also used for shaft systems with journal bearings, to control the approach to an operational regime with insufficient lubrication. Experiments [34] show that, near such a regime with intensive spalling, when scuffing is imminent, high-level pulses, or excursions, usually appear in the vibrational signal of a mechanism. In rotors with journal bearings, self-oscillation may be excited by the nonconservative fluid-film forces in the bearings. This self-oscillation, usually of frequency $\omega_s/2$, is identified by the methods of Chapter 8 (when its amplitude is not too high for continuous shaft operation, for some time). Methods for estimating the stability margin from vibration data (Chapter 7) may also be applied if the external random excitation is enough to generate a measurable subcritical response signal.

In conclusion, we stress that we have not tried to provide a complete survey of diagnostic methods for various types of rotating machinery. That would require a separate book devoted to this advanced field of application of mechanical signature analysis. In this chapter, we have outlined and illustrated the basic procedure, which seems to be common to many kinds of rotating machinery; namely, establishing a correspondence between the measured-response-spectral peaks and the characteristic rotational frequencies, and thus correlating the peaks with possible defects or faults. Similar examples of response-spectra interpretation in other types of rotating machine, particularly gearboxes, are given in [34,45,53]; in certain cases, cepstral analysis is useful for separating the sets of characteristic rotational frequencies and their multiples [34].

References

1. Alexandrov, A.A., Barkov, A.V., Barkova, N.A. & Schafransky, V.A., 'Vibrations and Vibrational Diagnostics of Ship Equipment' (in Russian). Sudostrojenie, Leningrad, 1986
2. Aström, K.J., 'Lectures on the identification problem – the least square method', Rep. Lund. Inst. Technol. Div. Automatic Control. Sweden, 1968
3. Balitsky, F.Ya., Ivanova, M.A., Sokolova, A.G., & Homyakov, E.I., 'Vibroacoustical Diagnostics of Originating Defects' (in Russian). Nauka, Moscow, 1984
4. Bendat, J.S., & Piersol, A.G., 'Random Data: Analysis and Measurement Procedures'. Wiley, 1971
5. Bendat, J.S., & Piersol, A.G., 'Engineering Applications of Correlation and Spectral Analysis'. Wiley, 1980
6. Bishop, R.E.D., 'Vibration'. Cambridge University Press, 1961
7. Bobrovnitsky, Ju.I., Genkin, M.D., & Dimentberg, M.F., 'Problems of acoustical diagnostics' (in Russian). *In* 'Vibroisolation Systems in Machines and Mechanisms' ('Vibroisoliruyuschie systemy v maschinah i mechanismah'). Nauka, Moscow, 1977
8. Bobrovnitsky, Ju.I., & Tjutekin, V.V., 'Energy properties of composite waveguides' (in Russian), *Acoust. Mag.*,1986, **32**, (5)
9. Bolotin, V.V., 'Nonconservative Problems in the Theory of Elastic Stability'. Pergamon, Oxford, 1963
10. Bolotin, V.V., 'Random Vibration of Elastic Systems'. Martinus Rijhoff, The Hague, 1984
11. Braun, S. (Ed.), 'Mechanical Signature Analysis, Theory and Applications'. Academic Press, 1986
12. Cawley, P. 'The reduction of bias error in transfer function estimates using FFT-analyzers', *J. Vibration, Acoustics, Stress & Reliability,* 1984, (1)

13. Cempel, C., 'VA diagnostics of machinery: the goals and the solution methods', Proc. XV Int. Conf. on Dynamics of Machines. Karl Marx Stadt, 1986
14. Chahine, G., & Courbier P., 'Noise and erosion of self-resonating cavitating jets', *J. Fluid Eng.,* 1987, (4)
15. Chang, P.C., Lin, A., Secor, G.A., & Su, K.S., 'Determination of the .pulse wave velocity by a filtered cross-correlation technique', *J. Biomech.,*1971, **4**, (6)
16. Connors, H.J., Jun., 'Fluidelastic vibration of tube arrays excited by cross flow'. *In* 'Flow-Induced Vibration in Heat Exchangers'. ASME, 1970
17. Cramer, H., & Leadbetter, M., 'Stationary and Related Stochastic Processes'. J. Wiley, 1967
18. Crandall, S.H., 'The role of damping in vibration theory', *J. Sound & Vib.,*1970, **II**, (1).
19. Crandall, S.H., & Mark, W., 'Random Vibrations in Mechanical Systems'. Academic Press, New York, 1963
20. Dimaragonas, A.D., & Paipetis, S.A., 'Analytical Methods in Rotor Dynamics'. Appl. Science Publishers, London, 1983
21. Dimentberg, M.F., 'Statistical Dynamics of Nonlinear and Time-Varying Systems'. Research Studies Press, 1988
22. Ewins, D.J., 'Modal Testing: Theory and Practice'. Research Studies Press, 1984
23. Frolov, K.V., 'Vibration: Friend or Foe?' (in Russian). Nauka, Moscow, 1984
24. Fu, K.S., 'Sequential Methods in Pattern Recognition and Machine Learning'. Academic Press, New York, 1968
25. Gersch, W., Brotherton, T., & Braun, S., 'Nearest neighbour-time series analysis classification of faults in rotating machinery', *J. Vib. Acoust. Stress & Reliability,* 1983, (2)
26. Goff, K., 'An application of correlation techniques to some acoustic measurements', *J. Acoust. Soc. Am.,* 1955, **27**, (2)
27. Hamming, R.W., 'Digital Filters'. Prentice-Hall, 1977
28. Hammond, J.K., 'Data analysis' . *In* 'Noise and Vibration'. (R.G. White & J.G. Walker, Eds.). Ellis Horwood, 1982
29. Harkievitch, A.A., 'Spectra and Analysis' (in Russian). Moscow, Gos. Izdat. Techniko-teor. Lit., 1953
30. Holmes, P.J., 'The experimental characterization of wave propagation systems', *J. Sound & Vib.,*1974, **35**, (2)
31. Jenkins, G.M., & Watts, D.G., 'Spectral Analysis and its Applications'. Holden-Day, 1969

32. Kryter, R.C., Robinson, J.C., & Thie, J.A., 'USA experience with in-service monitoring of core barrel motion in PWRs using ex-core neutron detectors'. *In* 'Vibration in Nuclear Power Plants'. Proc. Int. Conf. London, British Nuclear Energy Society, 1979
33. Lin, Y.K., Fujimori, Y., & Ariaratnam, S.T., 'Rotor blade stability in turbulent flows', *AIAA J.,* 1979, **17**, (6) (part 1), (7) (part 2)
34. Lyon, R.H., 'Machinery Noise and Diagnostics'. Boston, Butterworth, 1987
35. Max, J., 'Méthodes et Techniques de Traitement du Signal et Applications aux Mesures Physiques. Masson, 1981
36. Menyailov, A.I., 'Vibroimpact motions of control rods in control and safety systems of nuclear reactors' (in Russian), *Maschinovednie,* 1982, (6)
37. Mote, C.D., Jun., & Holoyen, S., 'Confirmation of the critical speed stability theory for symmetrical circular saws', *J. Eng. Ind.,*1975, (3)
38. Nigul, U.K., 'Nonlinear Acoustical Diagnostics' (in Russian). Sudostrojenie, Leningrad, 1981
39. Novikov, A.K., 'Statistical Measurements in Ship Acoustics' (in Russian). Sudostrojenie, Leningrad, 1985
40. Pal'mov, V.A., 'Integral methods for complex dynamical systems analysis' (in Russian), *Usp. Mekh.,*1979, **2**, (4)
41. Panovko, Ya.G., & Gubanova , I.I., 'Stability and Vibrations of Elastic Systems' (in Russian). Nauka, Moscow, 1967
42. Powell, R.E., & Seering, W., 'Multichannel structural inverse filtering', *J. Vib. Acoust. Stress & Reliability in Design,*1984, (1)
43. Ruhlin, C.L., Watson, Judith J., Ricketts, R.H., & Dagget, R.V., Jun., 'Evaluation of four subcritical response methods for on-line prediction of flutter onset in wind tunnel tests', *J. Aircr.,* 1983, **20**, (10)
44. Shahrivar, F., & Bouwkamp, J.G., 'Damage detection in offshore platforms using vibration information', *J. Energy Resources Technol.,* 1986, (2)
45. Stewart, R.M., 'Application of signal processing techniques to machinery health monitoring'. *In* 'Noise and Vibration'. (R.G. White & J.G. Walker, Eds.). Ellis Horwood, 1982
46. Sveshnikov, A.A., 'Applied Methods of the Theory of Random Functions'. Pergamon, New York, 1966
47. Tedrick, R.N., 'Effect of combustion instability on rocket-engine noise sources and levels', *J. Acoust. Soc. Am.,* 1965, **38**, (5), p. 921
48. Tikhonov, A.N., & Arsenin, V.Ya., 'Methods for Solving Ill-Posed Problems' (in Russian). Nauka, Moscow, 1974

49. Tsyfansky, S.L., Ojiganov, V.M., Milov, A.B., & Nevsky, Yu.N., 'On a certain method for searching damages in aircraft wings, based on the analysis of nonlinear oscillations' (in Russian). *In* 'Topics of Dynamics and Strength' ('Voprosy dinamiki i prochnosti'), (39), Riga, 1981
50. Vandiver, J.K., Dunwoody, A.B., Campbell, R.B., & Cook, M.F., 'A mathematical basis for the random decrement vibration signature analysis technique', *J. Mech. Des.,* 1982, (2)
51. Wedig, W. 'The integration of nonlinear stochastic systems with application to damage and ambiguity identification', *ZAMM,* 1981, **61**, (1)
52. White, R.G., 'Vibration testing'. *In* 'Noise and Vibration'. (R.G. White & J.G. Walker, Eds.). Ellis Horwood, 1982
53. Yavlensky, K.V., & Yavlensky, A.K., 'Vibrational Diagnostics and Quality Prediction for Mechanical Systems' (in Russian). Maschinostroenie, Leningrad, 1983
54. Zadeh, L., & Desoer, C., 'Linear Systems Theory'. McGraw-Hill, New York, 1963
55. Zhuravlev, V.Ph., & Bal'mont, V.B., 'Mechanics of Ball Bearings for Gyros' (in Russian). Mashinostroenie, Moscow, 1986

Index